T5-CAO-611

A series of student texts in

CONTEMPORARY BIOLOGY

The Life and Organization of Birds

W. B. Yapp
O.B.E., M.A.

Senior Lecturer in Zoology, University of Birmingham

American Elsevier Publishing Company, Inc.
New York

First published 1970

American Elsevier Publishing Company, Inc.
52 Vanderbilt Avenue, New York, N.Y. 10017

First published in Great Britain by
Edward Arnold (Publishers) Ltd.

Standard Book Number (Cloth edition): 444–19635–8
Standard Book Number (Paper edition): 444–19644–7
Library of Congress Catalog Card Number: 74–130965

Printed in Great Britain by
William Clowes and Sons, Limited, London and Beccles

Preface

When Hans Gadow ceased to be active, about the time of the first World War, ornithology ceased to be respectable among English zoologists. Since then it has slowly crept up in the scientific peck-order, but even now few Britons (in contrast to Americans) who would be recognized as interested in zoology as well as in birds attend the ornithological congresses. In writing this book I have tried to show zoologists that birds are interesting in their own right, as animals and as vertebrates, and that a biology based on the frog and the mammal is deficient. I hope that most of the book is written in language simple enough for the bird-watcher also to learn something from it. Some necessary technical terms are explained in Appendix 2.

I have not been able to cover everything, and in selecting topics for development I have chosen those that interest me, or that seem to me to be specially important for the main theme of the book. Other authors would have chosen differently, but I hope that my picture of birds as animals is reasonably fair, and is one to which students may add but from which they will not have to take much away.

No one is more aware than I am that I have not read all that has been written about birds, and that my field experience is limited to three continents and a few hundred species. My bibliography includes only books and articles from which I have quoted facts or that I have found stimulating for their opinions or discussion. Statements of fact not annotated, if not based on personal knowledge, are taken from one or more of the standard books listed in the first part of the bibliography.

I hope that my readers will enjoy the book as much as I have enjoyed writing it. I am grateful to Professor Barrington, the General Editor of the series, for suggesting it to me; his comments on the manuscript have been most helpful.

Church End
1970.

W.B.Y.

Acknowledgements

I am grateful to all those who have allowed me to reproduce figures or tables. Dr. David Snow, of the Bird Room of the British Museum (Natural History) and Mr. Sewell of the Birmingham Natural History Museum have given me facilities for study and lent me specimens; I thank them and their staff for all their help. The patience and assistance of Mr. Owen Harry and other members of the staff of the Department of Zoology and Comparative Physiology of the University of Birmingham have greatly lightened my labours.

W.B.Y.

Table of Contents

Preface v

1 REPTILES WITH FEATHERS 1

1.1 ANCESTRY 1

1.2 REPTILIAN FEATURES OF MODERN BIRDS 5

1.21 Eggs 5
1.22 Exoskeleton 8
1.23 Kidney 9
1.24 Blood system 11
1.25 Skull 12

2 FLIGHT 13

2.1 THE SKELETON 13

2.11 Forelimb 13
2.12 Shoulder girdle and sternum 14
2.13 Bipedality 17
2.14 Vertebrae and ribs 19
2.15 Pneumatization 21

2.2 MUSCLES 22

2.3 FEATHERS 23

2.4 THE MECHANICS OF FLIGHT 25

2.41 Flapping flight 26
2.42 Gliding and soaring 27
2.43 The slotted wing 28
2.44 The speed of flight 30

2.5 MUSCULAR CONTRACTION 30

2.6 FEATURES ASSOCIATED WITH FLIGHT 30

2.61 Temperature control 32
2.62 Colour 32
2.63 Moult 33
2.64 Respiratory system 35
2.641 Breathing 35
2.65 The senses 37
2.651 Sight 37
2.652 Hearing 37
2.653 Balance 38
2.66 The brain 38

3 CLASSIFICATION AND ADAPTIVE RADIATION 40

3.1 MODERN TAXONOMY 42

3.2 ADAPTIVE RADIATION 44

3.21 Flightless birds 45
3.211 The Ratitae 46
3.22 Swimming and diving birds 49
3.23 Typical birds 55

3.3 SPECIES AND SUBSPECIES 58

4 PHYSIOLOGY 65

4.1 NUTRITION 65

4.11 Food 65
4.12 The alimentary canal 66
4.13 Digestion 67

4.2 METABOLISM 70

4.21 Biochemistry of carbon 70
4.22 Biochemistry of nitrogen 71
4.23 The kidney 71
4.24 Production of energy 72
4.25 The physiology of diving 74

4.3 TEMPERATURE CONTROL 75

4.31 Control of loss of heat 76
4.32 Control of production of heat 77
4.33 Ontogeny of temperature control 80
4.34 Torpidity 81

4.4 NERVOUS SYSTEM AND SENSE ORGANS 81

4.41 Nervous system 82
4.42 Simple sense organs 84

4.43 Chemical senses 85
4.431 Taste 85
4.432 Smell 86
4.44 Sight 87
4.45 Hearing 89

5 THE ENDOCRINE CONTROL OF REPRODUCTION 92
5.1 THE FACTS OF REPRODUCTION 92
5.11 The gonads 92
5.12 Breeding seasons 94
5.13 Hormones connected with reproduction 96
5.131 The adenohypophysis 96
5.132 The sex hormones 97
5.2 LIGHT AND THE PITUITARY 97
5.21 Experimental induction of breeding 97
5.22 The breeding cycle 101
5.3 SOME OTHER HORMONAL RELATIONSHIPS 104
5.31 Plumage 104
5.32 Other epidermal structures 106
5.33 The egg shell 107
5.34 Behaviour 107

6 THE HIGHER LIFE 109
6.1 COMPARISON OF BIRDS WITH MAMMALS 109
6.11 Instinct and intelligence 109
6.12 Innate behaviour 111
6.13 Learning 111
6.14 Instinctive or intelligent? 113
6.2 REPRODUCTIVE BEHAVIOUR 114
6.21 Pair formation 114
6.22 Nests 117
6.221 Choice of site 118
6.222 Nest building 120
6.223 Nest sanitation 122
6.23 Brooding 123
6.24 Non-nesters 124

7 MAINTENANCE ACTIVITIES OF REPRODUCTION 126
7.1 SONG 126
7.11 The structure of song 126
7.12 The function of song 132
7.13 Song dialects 135
7.14 Learning to sing 136

7.2 CALL-NOTES 139
7.3 TERRITORY 139
7.31 The function of territory 142
7.4 MIGRATION 145
7.41 The advantages of migration 149
7.42 The stimulus for migration 153
7.43 The physiology of migration 154
7.44 Navigation 156
8 OTHER COMPLEX BEHAVIOUR 163
8.1 FOOD-SEEKING 163
8.11 Storage of food 165
8.12 Tool-using 165
8.2 FLOCKING 166
8.21 Colonial nesting 166
8.22 Winter flocks 167
8.23 Roosting 168
8.24 Behaviour in flocks 169
8.3 MOBBING 169
8.4 SELF-STIMULATION 171
9 DISTRIBUTION AND DISPERSION 172
9.1 STABILITY OF NUMBERS 172
9.11 Some decreases 175
9.12 Some increases 181
9.2 CLUTCH-SIZE 183
9.3 MORTALITY 187
9.31 Death from starvation 187
9.32 Death from predation 189
9.4 LIMITATION OF NUMBERS BY SPACE OCCUPANCY 190
9.5 HABITAT SELECTION 193
9.51 The influence of other species 196
9.52 Changing habitat 198
9.6 POPULATION CONTROL: SUMMARY 199
Appendix 1 Classification of birds and list of species 201
Appendix 2 Glossary 210
References 215
Index 233

I

Reptiles with Feathers

Birds may be reptiles with feathers, but to say this, though true, is no more complete than to say that man is a two-legged monkey. Whatever the anatomical and physiological similarities between man and ape, man has powers and a mode of life different in kind from those of other mammals. So with the birds; their structure and physiology may be close to those of reptiles, but only a zoologist who never saw them alive would want to put the two groups in the same class.

1.1 ANCESTRY[158]

The skeleton of birds has long been known to be similar in many ways to that of the dinosaurs now called Ornithischia, one of the groups of diapsid reptiles or Archosauria that flourished in the Triassic. The heart and blood system closely resemble those of crocodiles, which are themselves diapsid and also Archosauria. It is therefore in this group that we may hope to find the ancestors of birds. The Ornithischia are too specialized to be on the direct line; for example they have no clavicles, and some of their similarities to birds appear to be convergences. Another Archosauran group, called Pseudosuchia (Fig. 1.1), also from the Triassic, is primitive enough for some of its members, or at least their near ancestors, to be the animals from which both birds and dinosaurs were derived; (the Pseudosuchia are in some classifications a subdivision of the Thecodontia; in others the two names are alternatives). The separation of birds from the other reptilian lines perhaps took place in the Permian.

Unfortunately the fossil record is poor. Bird bones are thin, and easily destroyed, and although some thousands of bird fossils have been described most are fragmentary and few are old. There is, however, one early

Fig. 1.1 Restoration of *Ornithosuchus*, a pseudosuchian. (From a postcard, British Museum, Natural History.)

Fig. 1.2 *Archaeopteryx lithographica*, reconstruction. (From W. E. Swinton, 1958, *Fossil Birds*, Plate 1, British Museum, Natural History.)

bird of which three specimens are known (or four if the remains of a single feather are counted). Two are fairly complete. All can be put in the same species, *Archaeopteryx lithographica*[80] (Fig. 1.2). The specimens were found in two quarries about 10 miles apart, and at different levels in the limestone, so that the species must have been well established and must have persisted for some time. The period was the Jurassic, which is right for the creature to be a link between the Pseudosuchia and modern birds. It is also a link in structure, for while the skeleton (Fig. 1.3) is clearly

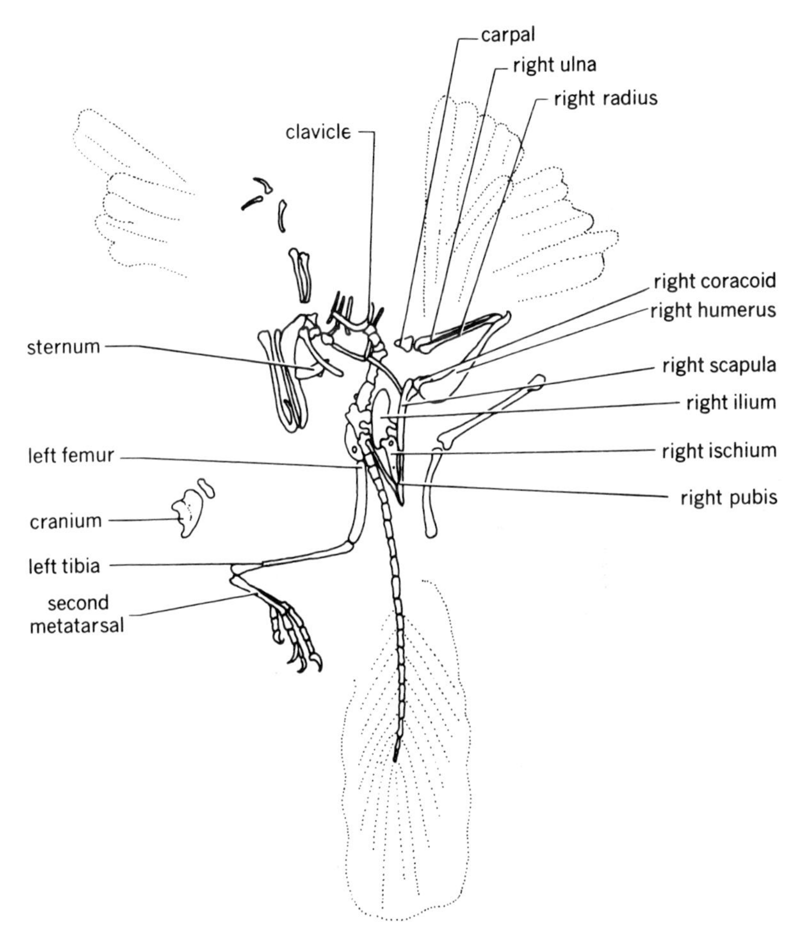

Fig. 1.3 *Archaeopteryx lithographica*, British Museum specimen, ×1/3. (From W. B. Yapp, 1965, *Vertebrates: their structure and life*, Fig. 6.10, Oxford University Press, New York.)

reptilian, it has a few features that recall birds; for example the two clavicles are fused at one end to form a wishbone; the pubis is directed backwards; the first toe is opposable; and there appears to be fusion in the ankle and foot to form four tarsometatarsals, and in the wrist to form a single carpometacarpal corresponding to the third digit; (tarsometatarsus and carpometacarpus, characteristic of birds, were present in some dinosaurs also). Purely reptilian features were the long tail, the amphicoelous or amphiplatyan vertebrae, the peg-like teeth and the simple brain with cylindrical cerebral hemispheres and unexpanded cerebellum.

The feature that made Owen, who first described it in 1863,[281] and almost everyone who has studied it since, class *Archaeopteryx* as a bird, is its feathers. These are shown in the three main specimens only as impressions in the rock, but the grain of this is so fine that under the microscope detailed structure down to barbules and hamuli can be seen, corresponding exactly to that of modern birds (Chapter 2). Some of the feathers are arranged like the leaflets of a fern along the axis of the tail, and others lie detached from the body but in such a position that they might well have made a wing consisting of primary and secondary quills. In the Berlin specimen there are smaller feathers, or coverts and contour feathers, in various parts of the body. The skeleton suggests that the pectoral muscles were small, so that if the creature flew at all it must have done so at best by feeble flapping; possibly it only glided in the manner of flying squirrels. The claws on the hand (not visible in the British Museum specimen) and the opposable first toe suggest that the animal was arboreal. Altogether, it seems to be the nearest to a perfect example of a missing link that has ever been found, showing some of the characters of its ancestors, some of those of its descendants, and living an intermediate sort of life.

Although *Archaeopteryx* was well established and probably lived throughout Europe at least (a downy feather of much the same age, that may belong to the same species, has been found in Spain), nothing is known of its immediate descendants. The next fossils of birds are Cretaceous, and are scattered from Rumania to Wyoming and Sweden to Chile, some being found in the Cambridge Greensand. They are of different types, but tell us little or nothing of how birds evolved. All appear to have been marine, but this does not show that birds had swimming ancestors; probably aquatic creatures are the ones most likely to be fossilized. The best-known genera, *Ichthyornis* and *Hesperornis*, both from Kansas, are usually figured with teeth, but many palaeontologists now hold that the alleged jaws of *Ichthyornis* belong to associated reptilian skeletons, and the same is possibly or probably true of *Hesperornis* also. While *Ichthyornis* had a big keel on the sternum and so may have flown,

Hesperornis had no keel, and only the stump of a forelimb, so that it must have been flightless.[354]

In the Eocene period and later there are many bird fossils, mostly fragmentary, and many related more or less clearly to existing families. It seems likely, therefore, that birds as we know them today were established before the end of the Cretaceous. They evolved, in fact, at the same time as the flowering plants, the mammals, and the higher orders of insects.

1.2 REPTILIAN FEATURES OF MODERN BIRDS

In addition to many points in the skeleton, some of which are mentioned in the next chapter, birds resemble reptiles in some other striking ways. One of the most obvious is that they lay eggs. So do fish and frogs, and indeed most invertebrates, but the eggs of reptiles and birds agree, and differ from those of all other animals except the monotreme mammals, in being large and in having a thick shell.

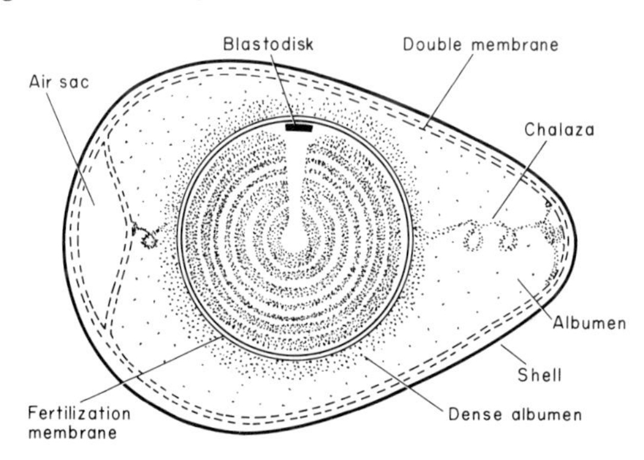

Fig. 1.4 Diagrammatic sagittal section of a hen's egg. (After W. B. Yapp (ed), 1963, *Borradaile's Manual of Zoology*, Fig. 28.19, Oxford University Press, London.)

1.21 Eggs

The structure of a hen's egg is shown in Fig. 1.4. Like that of a reptile, it is described in technical language as telolecithal, because the ovum contains a large amount of yolk, at one pole of which floats the germinal disk, formed of cytoplasm with its nucleus. More strictly, the granules of yolk, being slightly denser than cytoplasm, sink to the bottom of the spherical cell. Yolk consists largely of the phospholipid lecithin (sometimes regarded as 'yolk proper') much of which is attached to the protein vitellin to form lecithovitellin. Fat is also present, and the gross analysis of the solids in the yolk of a hen's egg is more than 60 per cent fatty substances, and more than 30 per cent protein. The yolk is arranged in alternating layers, sometimes called white and yellow, the former containing less fat and so less of the fat-soluble pigment carotene to which the yellow colour is due. A column of white yolk stretches from the centre to the germinal disk or blastodisk. The limit of the ovum, or egg proper, is the vitelline membrane that surrounds the 'yolk' as that word is used in cookery. In biological language, the vitelline membrane surrounds the yolk plus the cytoplasm and nucleus. The cytoplasmic vitelline membrane is transformed at fertilization into the fertilization membrane. The tough sheath that surrounds the yolk (well seen in a boiled egg) consists of cellular and fibrous material derived from other cells of the ovarian follicle, and is therefore not strictly a vitelline membrane or part of the egg cell.

If fertilization occurs, it does so high in the oviduct, and division of the zygote begins almost at once. As the ovum passes down the oviduct, material is added which is no part of the female gamete, but which is an important part of the egg of ordinary language. First is the white (if the alternative name albumen is used, the spelling should be carefully noted). It is arranged in four layers, and consists mostly of a solution of the protein albumin, but other proteins are present and there are fibres of a glycoprotein called ovomucoid, similar in chemical structure to the mucins of mammalian secretions. The innermost layer of the albumen consists of a dense rope of these, the chalaza, running from one end of the egg to the other, and spreading out in the middle to surround and be attached to the membrane of the yolk. It can rotate within the more liquid surrounding layer of white, so that when the egg is turned the lighter blastodisk can turn also and so come uppermost. The white is surrounded by two shell membranes, which consist mostly of keratin, and outside these is the shell itself, at least 94 per cent of which is calcium carbonate. The two membranes separate to form an air space at the blunt end when the egg is laid. In the monotremes, and in most reptiles, the shell is leathery, consisting of keratin without calcium carbonate.

Both yolk and egg-white supply food for the developing embryo, the necessary protein coming from the two sources in equal amounts. In the later stages of incubation some of the shell is dissolved and the calcium carbonate is used to build the developing bone. The shell also supplies a little protein. The egg-white supplies water and minerals, while vitamins come both from yolk and white.[304]

Birds' eggs differ from those of reptiles in that many of them are coloured. Several pigments have been described, but all seem to be porphyrins or their relations the bile pigments, all of which can be formed, and probably are in nature, by the breakdown of haemoglobin.

Since the general acceptance of natural selection as the only mode of evolution many attempts have been made to explain the colours of birds' eggs by this hypothesis. It has been repeatedly said that the eggs of birds that nest in holes and caves are white, and that only when they are incubated in the dark could such conspicuous objects survive. Further, it is said that coloured eggs have been selected because they are cryptic, or sometimes because they are conspicuous but unpalatable. The eggs of many birds that nest on the ground are certainly well-concealed, and the establishment of their colours by natural selection would seem reasonable. But there are too many exceptions for this to be a general explanation. Reptiles' eggs, though white, survive, and several orders of birds—petrels, parrots, pigeons, owls, for example—lay white eggs irrespective of where they nest. The pigments of birds' eggs are waste products, and it seems most likely that their deposition was originally an accident, but that colours have occasionally been selected because of their value in concealment.

The large, yolky, thick-shelled egg of reptiles and birds probably evolved because of the advantage that it gave in a dry climate by aiding the conservation of water. The shell is pierced by pores, through which gases can pass, but they are too fine to allow the passage of much liquid except under considerable pressure. The egg is to a high degree shut off from its surroundings and is called cleidoic. In the reptiles that have been studied about a third of the water required enters the egg during its development, whereas in the bird the water produced by metabolism, added to that already present, makes a surplus, and a small amount evaporates. Nitrogenous waste accumulates inside the egg; in the birds and the lizards and snakes ammonia reacts with other substances to produce uric acid, a relatively insoluble and so harmless compound. One may guess that the pseudosuchians and dinosaurs also produced it.

Water is saved because if the more usual soluble waste products, urea and ammonia, were produced, they would need to be removed in solution. It is saved also because while in the early development of fish and amphibians up to 90 per cent of the material oxidized to provide energy is protein, in the chick over 90 per cent consists of fatty substances, which produce more water per unit mass.

Although a few reptiles, such as pythons and crocodiles, guard and protect their eggs, the majority leave them to hatch in ordinary environmental conditions.[68] Birds are very different; nearly all incubate their eggs by the heat of their bodies, and the majority make more or less elaborate nests to make this possible. Birds' eggs will not develop at atmospheric temperature, and in many species there is a change-over from one parent to the other as incubator, or the eggs may be covered by down or plant material when they are left, so that conditions are kept more or less constant. The nesting habits of birds are possible only because the cleidoic egg was already in existence. Their advantage is that they lead to a quicker development, so shortening the period of vulnerability, while the shelter that the eggs often receive possibly reduces the risk of destruction. Against this, the sitting bird becomes an easier prey for predators; black-headed gulls, especially on the edge of the colony, have been shown to be an easy prey for foxes.[282a]

1.22 Exoskeleton

The feet and lower parts of the legs of most birds are covered either with distinct scales or with a continuous thick sheath. In both types of structure the dermis is thickened, and outside it is a thick patch of keratin formed in the epidermis. This is the structure of a reptilian scale, and there

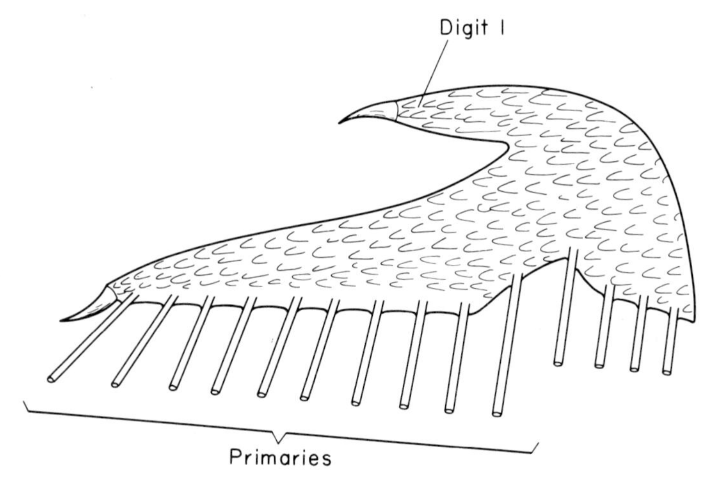

Fig. 1.5 Wing of a young hoatzin, ×1.0.

seems no reason to doubt that the scales of birds and reptiles are homologous. We know nothing of the evolution of feathers from scales, but intermediates between scale and feather have been found on the legs of ostriches and fowls. Reptilian scales are normally shed and replaced, and there is some evidence that the same happens in birds.

Claws, in reptiles, birds and mammals, have the same basic structure as scales. They occur regularly on the toes of birds, and on the first digit of the forelimbs in ostriches, hens, geese, game-birds and birds of prey; in some they are present on the second digit also. They generally disappear or become insignificant after hatching, and the only instance of their use in modern birds seems to be in the hoatzin (*Opisthocomus hoazin*, Fig. 1.5) of South America, where in the young they have functional muscles, so that the bird clambers about the trees with their aid. Traces of wing claws occasionally turn up in other species; they are clearly vestigial structures.

1.23 Kidney

The basic structure of the kidney is the same in all vertebrates. It is a collection of nephrons, which are coelomoducts, each of which has an inner end called Bowman's capsule, enclosing a small portion of coelom, and then runs as a tubule to collecting ducts and so to the hind gut, or more rarely directly to the exterior. The capsule encloses a glomerulus, which is a knot of capillaries supplied with arterial blood, and in most vertebrates the tubules have a secondary supply from the renal portal system, which brings venous blood from the posterior part of the body.

Reptiles and most amphibians have kidneys that correspond closely to this general pattern, with no marked specialization or loss of any one part. That of reptiles differs embryologically in being a metanephros instead of a mesonephros but this does not affect the structure of the nephron. The kidneys of birds resemble those of reptiles, but show two specializations. The first is that the tubules are relatively longer, and in many of them there is inserted in the middle part a U-shaped piece called the loop of Henle; in this they resemble the condition in mammals, to which the name properly applies, but they must have evolved independently, and there has undoubtedly been convergence. The other point is that although there is, in gross terms, a renal portal system, and phenol red injected into one leg of a fowl is excreted by the kidney of the same side, there is no certainty as to how far the system is functional in life.[338] The layout of the vessels, compared with that of a reptile, is shown in Fig. 1.6. Near the anterior end of the kidney there is a wide anastomosis between the renal portal vein and the renal vein, and it would seem impossible, when it is open, for more than a trickle of blood to pass through the capillaries round the tubules which are undoubtedly supplied by the branches of the renal portal. In some birds, however, there is a muscular constriction in the anastomosis, which can be made to contract by acetylcholine and to relax by adrenaline.[356] If it closes in life more blood would flow through the renal portal system, but even if this were so, the existence of the anastomosis, unknown in any reptile, suggests that the renal portal system is on the way out. There is none in mammals.

All the special points about the birds' kidneys serve the need to save water. Lizards and snakes excrete uric acid from the tubules, which is a parallel to the condition in birds, since it is unlikely that the common ancestors of the two groups had kidneys that worked in this way. It is unfortunate that the crocodiles, which in heart and skull are the nearest of

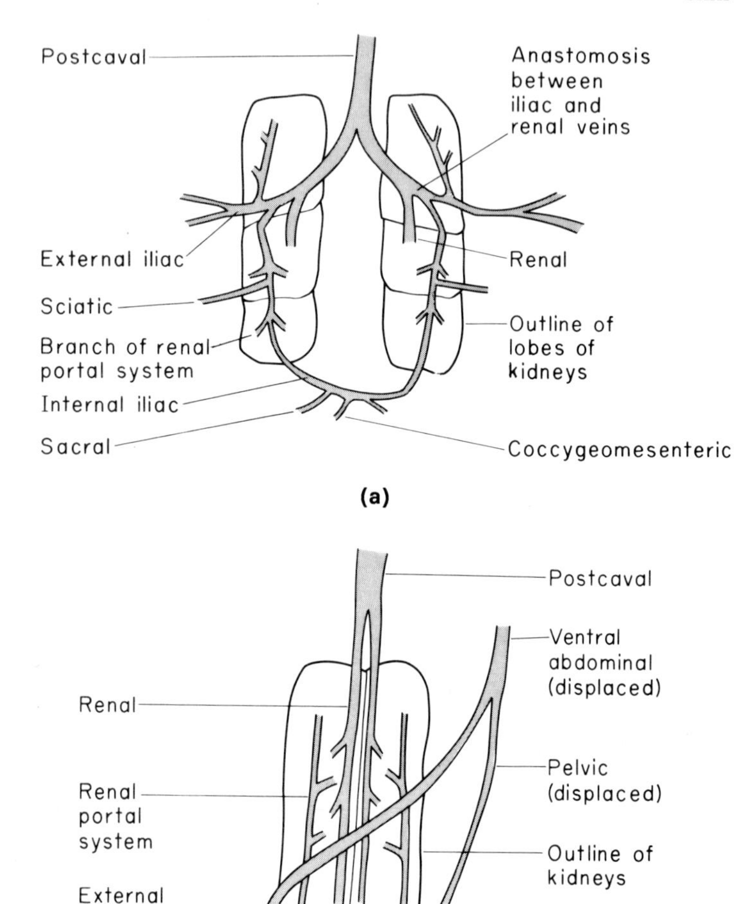

Fig. 1.6 (**a**) Renal portal system of a bird. (Based on I. Sperber, 1948, *Zool. Bidr. Upps.* **27**, 429.) (**b**) Renal portal system of a reptile.

living reptiles to the birds, are aquatic and so have different needs in excretion and throw no light on the history of the bird's kidney.

1.24 Blood system

Birds, like mammals, have a double circulation. The heart works as two pumps in parallel, the right side receiving de-oxygenated blood and sending it to the lungs, while the left side receives oxygenated blood from the lungs and sends it to the body generally. While in mammals the single systemic arch that arises from the left ventricle curves to the left side before it becomes the dorsal aorta, that of birds curves over to the right (Fig. 1.7a).

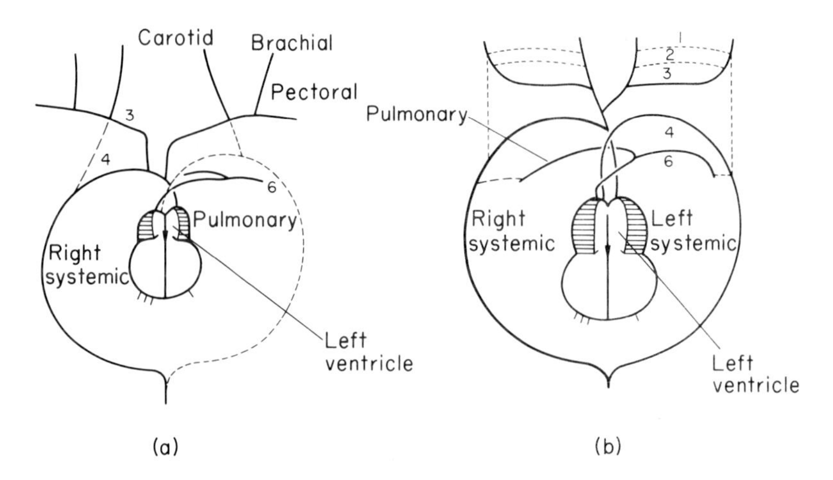

Fig. 1.7 Diagrams of the arterial arches (**a**) of a bird and (**b**) of a crocodile, ventral. The heart is shown flattened and pulled out, so that the auricles appear in their embryonic posterior position. Continuous lines show vessels that are always present, broken lines those sometimes present, and dotted lines those never present. (From W. B. Yapp, 1965, *Vertebrates: their structure and life*, Fig. 10.13, Oxford University Press, New York.)

This difference indicates independent evolution of the two types of heart. No reptile has anything approaching the mammalian condition, and we can only assume that stages in its evolution would have been found in the extinct therapsid reptiles which are believed to be the ancestors of mammals. The arrangement in the birds, however, can easily be derived from that in crocodiles (Fig. 1.7b). Here there are two systemic arches which cross over, that from the left ventricle curving dorsally to the right side, and that from the right ventricle to the left side. The vessel that goes to the left is much smaller than the other, and its suppression would give the arrangement found in the bird.

Another difference between the groups is that the septum between the ventricles is formed in different ways. While that in mammals is formed by the growth of a vertical ridge, in the same way as that of lizards and tortoises, that of birds is a different structure, formed by the coalescence of trabeculae, bars of muscle that cross the cavity of the ventricle in many reptiles and amphibians. A similar septum, originally horizontal, divides the ventricle of crocodiles. The heart and arteries of birds and crocodiles are therefore closely comparable.

1.25 Skull

It is difficult to see any detailed resemblance between the skull of an adult bird and that of any reptile, since all the sutures are obliterated. Formally, the bird's skull can be derived from the diapsid type. The arrangement of the bones in the lower jaw of a bird is almost purely reptilian (Fig. 1.8).

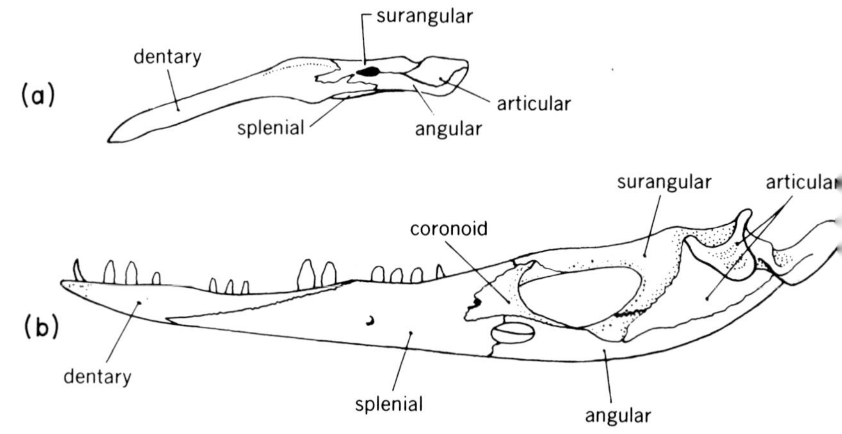

Fig. 1.8 Lower jaw of (**a**) barn owl, *Tyto alba*, outer aspect, ×1, and (**b**) *Alligator* sp., inner aspect, ×1/3. (From W. B. Yapp, 1965, *Vertebrates: their structure and life*, Fig. 16.23(c)/(d), Oxford University Press, New York.)

2

Flight

If it were not for the pterodactyls, a snap definition of the birds might be that they are flying reptiles. A few, indeed, do not fly, but all, except the moas (section 3.211), have wings; the possession of these, and of other features that are meaningless except as helps in flight, has convinced all but a very few zoologists that flightless birds all had flying ancestors. The only possibly primarily flightless bird was *Archaeopteryx*, which even if it had not quite achieved flight, must have been well on the way to doing so.

Books often list adaptations to flight in birds, but caution is needed in interpreting this phrase; it depends on what you mean by 'adaptation'. Some features are necessary for flight, for example wings, others merely assist it, for example feathers; some were probably selected because of the advantage they conferred in enabling birds to fly, others were probably selected for other reasons and were merely used when flight began. Some changes of each sort may have necessitated or brought with them changes in other parts of the body.

2.1 THE SKELETON

2.11 Forelimb

Flight is a means of locomotion, and locomotion in vertebrates depends chiefly on the skeletal and muscular systems, of which the first is the easier to study. It is clear from a comparison of bird, pterodactyl and bat (Fig. 2.1) that vertebrate flight does not rest upon any one pattern of skeletal structure. The properties that the skeleton of a wing must possess are ability to move up and down (for lift) and backwards and forwards (for propulsion), and some means of being put out of the way by folding or lying flat against the body. All these are shown, for example, by the human arm, and there seems no reason why a pentadactyl limb with a relatively unmodified skeleton, such as man possesses, could not perfectly well have been used for flight by the addition of the necessary surface. The peculiar structure of the bird's forelimb is not, therefore, an adaptation to flight in the first sense pointed out above. The chief differences from the standard pentadactyl pattern are the reduction in number and length of the digits and the simplification of the wrist, so that the proximal carpals are reduced, and the distal ones fused with the metacarpals and with each other to form a carpometacarpus (Fig. 2.2). None of these has any direct connection with flight.

2.12 Shoulder girdle and sternum

If the forelimb is to be moved up and down (instead of backwards and forwards as it is in most mammals), there must be some skeletal support

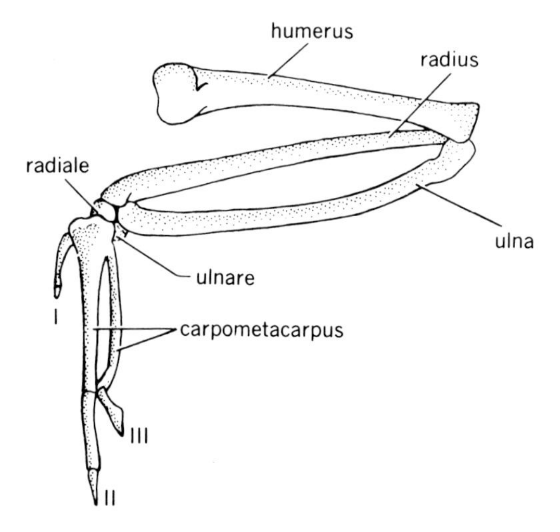

Fig. 2.1 (a)

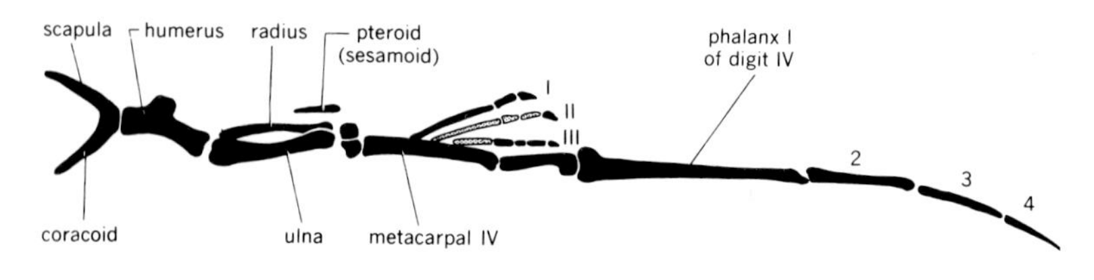

Fig. 2.1 (b)

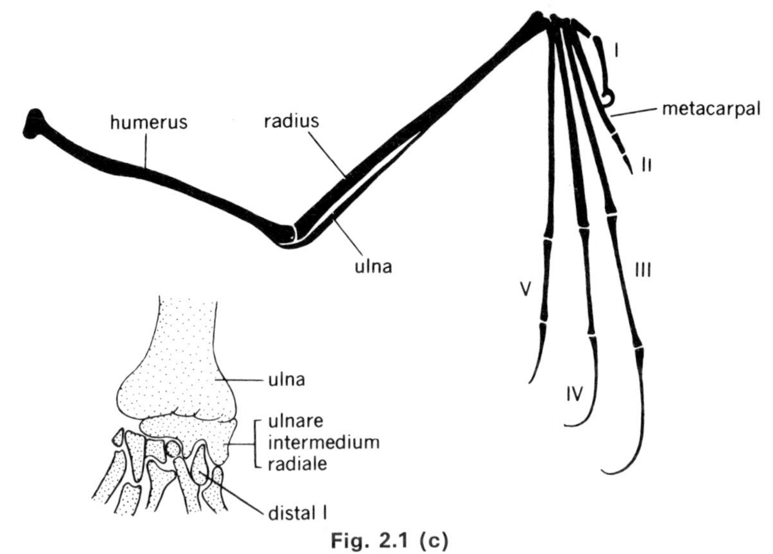

Fig. 2.1 (c)

Fig. 2.1 (**a**) Skeleton of the wing of a pigeon, *Columba livia*, ×1. (**b**) Skeleton of the wing of a pterodactyl, *Pteranodon* sp., ×1/12. The dotted bones are reconstructed. (**c**) Skeleton of the wing of a bat, *Pteropus* sp., ×1/2. Inset: detail of the carpus, ×2; there is one large fused proximal carpale, no centrale, and five separate distal carpals. (From W. B. Yapp, 1965, *Vertebrates: their structure and life*, Figures 6.6, 17.17 and 17.18, Oxford University Press, New York.)

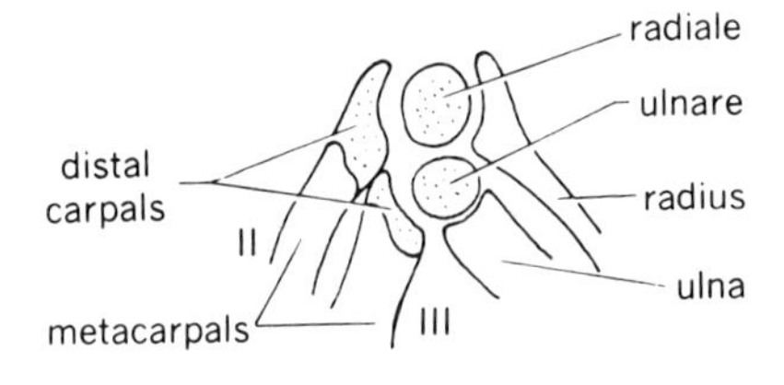

Fig. 2.2 The right wrist of a young house-martin (*Delichon urbica*) ventral, ×3. The distal carpals are not yet completely fused to the metacarpals. (From W. B. Yapp, 1965, *Vertebrates: their structure and life*, Fig. 6.5, Oxford University Press, New York.)

across the thorax to prevent its collapse when the muscles drawing the limb down contract. In man this is provided by the clavicles or collar bones, and in birds partly by these but mainly by the procoracoids (often incorrectly called coracoids). The clavicles are slender, and usually meet vertically in an interclavicle to form the wish-bone or furcula (Fig. 2.3). In some owls and others they do not unite, and in scattered species, including many parrots, they are reduced; they are absent from most ratites. The procoracoids are short bones and meet the sternum. The scapulae are characteristically sword-shaped and lie flat along the dorsal surface of the ribs. In frigate-birds (*Fregata*), which have very strong and graceful flight,

Fig. 2.3 Furcula and shoulder girdle of a peregrine, ×*c*. 0.5.

as well as in Ratitae, which are flightless, the scapulae and coracoids are fused.

Strong flight necessitates large muscles, and large muscles must have a big surface of bone for their origin. This is provided in most birds by a vertical extension of the sternum, the keel or carina (Fig. 2.4). Its relative

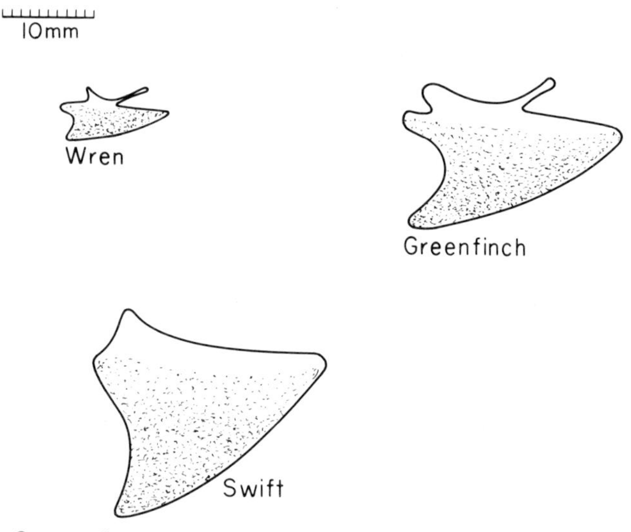

Fig. 2.4 Sterna of wren, greenfinch and swift to the same scale. Lengths of skull, occiput to frontonasal suture, are: wren 15 mm, greenfinch 20 mm, swift 21 mm.

size is in general proportional to the power of flight, so that it is very deep in swifts and humming-birds and almost absent from many rails and from

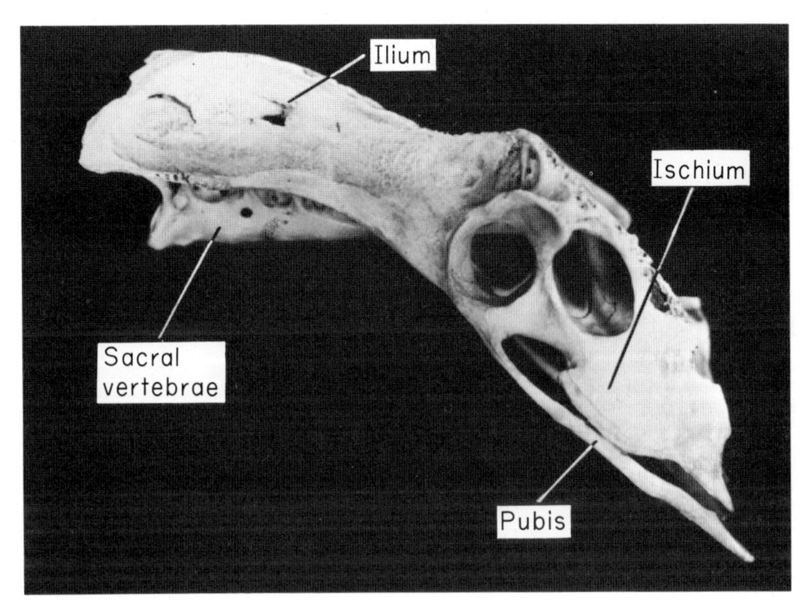

Fig. 2.5 Synsacrum of a duck, ×*c*. 1.0.

the flightless dodo (*Raphus cucullatus*) and kakapo (*Strigops habroptilus*) as well as from the Ratitae (Chapter 3).

2.13 Bipedality

Birds are greatly superior to bats and pterodactyls in supporting the wing entirely by the forelimbs, so that the hindlimbs are free and the animal can walk and run. Bipedality is as characteristic of birds as flight, or more so, since the flightless birds have all retained the habit of walking on two legs. That the loss of use of the forelimbs in running is no disadvantage is shown by its frequent occurrence in reptiles such as eosuchians and many dinosaurs, and in mammals such as kangaroos and jerboas. In these cases we must indeed assume that it was selected because it gave an advantage, probably in speed, and there are lizards and mice, normally quadrupedal, that rise on their hindlimbs when they are in a hurry. It seems likely, both from their origin from Pseudosuchia and from the freedom of their legs, that birds were bipedal before they flew; the structure of *Archaeopteryx* does not contradict this.

One would expect two legs that have to do the work of four to be relatively stronger, and this is true of most birds. The pelvic girdle is strong and the bones are fused with each other and with a number of vertebrae to form a synsacrum (Fig. 2.5). The absence of a ventral symphysis is

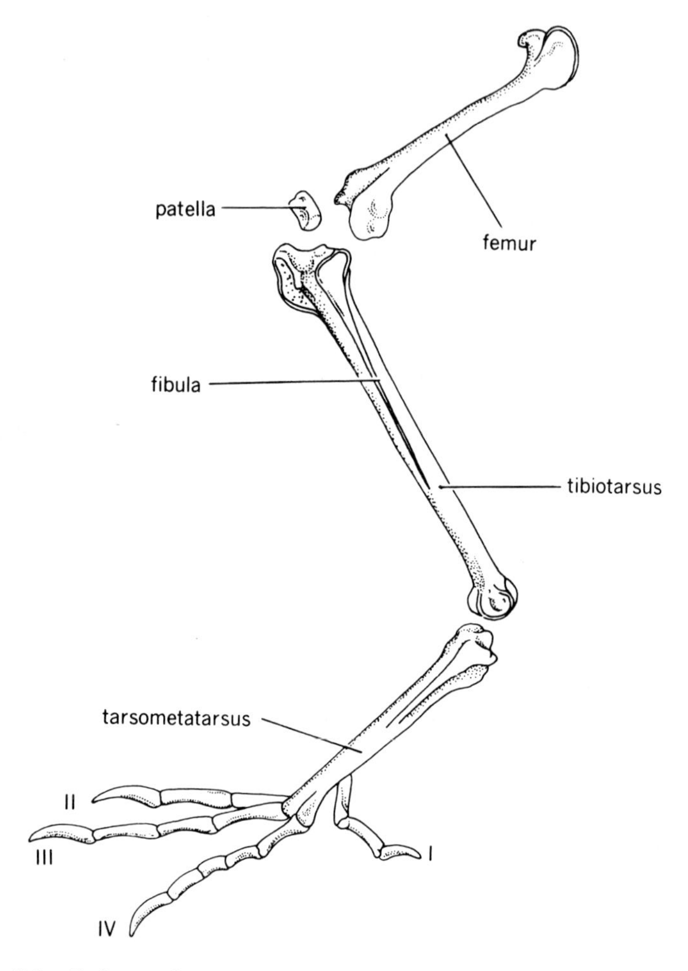

Fig. 2.6 Skeleton of the leg of a pigeon, *Columba livia*, ×4/3. (From W. B. Yapp, 1965, *Vertebrates: their structure and life*, Fig. 6.8, Oxford University Press, New York.)

probably connected with the large eggs that birds lay, since there needs to be a wide passage through which they can pass out. The ankle shows the same sort of changes as the wrist, but they have gone even further (Figs. 2.6 and 2.7). Not only are the distal tarsals fused with the metatarsals to form a tarsometatarsus, but the proximal tarsals are fused with the lower end of the tibia to form a tibiotarsus. These changes can reasonably be regarded as being valuable because they strengthen the limb for bipedal gait. The most likely explanation of the carpometacarpus is that it was produced by the same genes as produced the tarsometatarsus, and was originally of no selective value. This interpretation is supported by the probable occurrence of tarsometatarsus and carpometacarpus in pseudosuchians and in some bipedal dinosaurs such as *Iguanodon*. Later, the particular form of the bird's carpometacarpus may have been selected because of its value in flight.

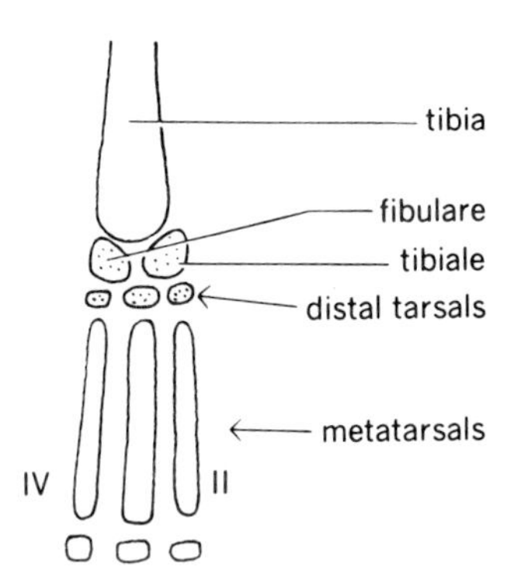

Fig. 2.7 The right ankle of a young house-martin (*Delichon urbica*) dorsal, ×3. The tibiale and fibulare are separate from the tibia, and the distal tarsals are separate from the metatarsals. (From W. B. Yapp, 1965, *Vertebrates: their structure and life*, Fig. 6.7, Oxford University Press, New York.)

The toes of birds are reduced to four, of which the first is normally directed backwards. The triple distal end of the tarsometatarsus, articulating with the second, third and fourth toes, is characteristic, but something similar is found in the jumping mouse, *Dipus*.

2.14 Vertebrae and ribs

Other features of the skeleton that have been claimed to be connected with flight are the heterocoelous vertebrae of the neck (Fig. 2.8a, b), the pygostyle (Fig. 2.8c), the compact thorax, and the uncinate processes of the ribs (Fig. 2.9a). The heterocoelous vertebrae confer great flexibility (all birds can twist their neck through 180°, and the wryneck (*Jynx*) can manage 360°), but probably have nothing directly to do with flight. Some birds do indeed turn their heads while flying to look to one side, but most fly looking straight ahead. Some of those with long necks used for catching fish, such as herons, fold their neck in flight in a way that makes turning it impossible. The flexible neck does aid the preening of the feathers in all parts of the body except the neck itself, and to that extent is of some value to flight.

Anyone who has made paper darts learns that the tail must not be too long, and there can be no doubt that the shortening of the caudal vertebrae

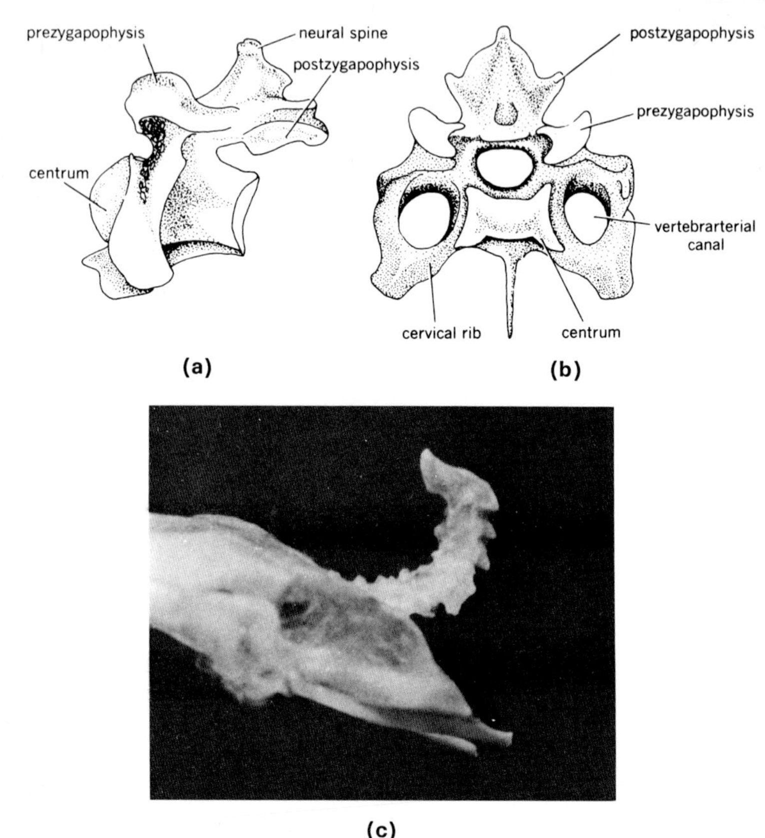

Fig. 2.8 Cervical vertebra of a turkey, *Meleagris gallopavo*, ×1.3, (**a**) lateral, and (**b**) anterior. (**c**) Pygostyle of a barn owl, *Tyto alba*. (a and b from W. B. Yapp, 1965, *Vertebrates: their structure and life*, Fig. 15.4, Oxford University Press, New York.)

to the pygostyle has assisted stability in the air. It is, however, going too far to say that because of its long tail *Archaeopteryx* could not have flown. The male whydahs (*Vidua*) of Africa, in their breeding dress with tail feathers several times the length of the body, do fly, although slowly and clumsily.

A compact thorax will perhaps help in flight by concentrating the mass, and uncinate processes will help in producing compactness. But they are present also in some reptiles such as crocodiles and *Sphenodon*, and occurred in the early amphibian *Eryops*. None of these ever flew, so that although the uncinate processes of birds may have been evolved in connection with flight (they were not present in *Archaeopteryx*) this is not

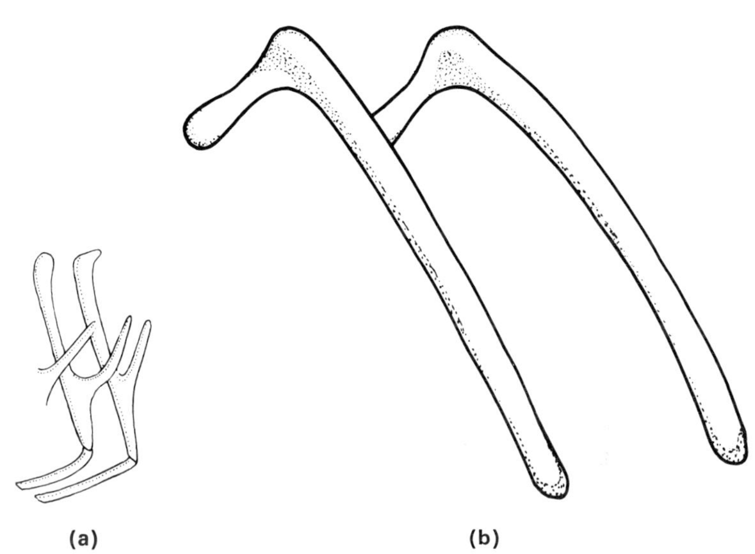

Fig. 2.9 (**a**) Fourth and fifth thoracic ribs of a barn owl, *Tyto alba*, ×2/3. (From W. B. Yapp, 1965, *Vertebrates: their structure and life*, Fig. 15.9, Oxford University Press, New York.) (**b**) Two ribs of a screamer, *Chauna torquata*, ×1.0.

certain. They are absent from the screamers (*Anhimae*), which can fly perfectly well (Fig. 2.9b).

2.15 Pneumatization

Many of the bones of birds contain air-spaces. Those of the skull develop by intrusions from the nasal cavity or the middle ear, those of the other bones in a similar way from the air-sacs. As the membranous walls of the cavities containing air grow into the bones the marrow and most of the trabeculae are destroyed. It is easy to suggest that the function of these pneumatic bones is to save weight and so assist in flight, but this is only true if there is no weakening of the bones. In general this appears to be the case. A hollow cylinder has a greater resistance to bending than a solid rod of the same total cross-section of material; hence the use of tubes in bicycles and other machines. In general pneumatized bird bones have a greater overall diameter than the ordinary bones of comparable species, suggesting that the disposition of the solid material is such as to give great strength for weight.

The distribution of pneumatization is not quite what might have been expected.[5,12] It is small in many diving birds, such as ducks and cormorants, where carrying unnecessary air below the surface would mean an

extra expenditure of energy, but there is also little in some good fliers such as swifts, game-birds and plovers. It is well developed in albatrosses and eagles, which as large flying birds have great need to save weight, but there is also much in the ratites. The air space in the femur of these can be explained as giving high strength-for-weight in the legs of large animals, but that in the humerus must have some other function or none at all. Air-spaces are present in the skull of crocodiles, and were probably present in the bones of pterodactyls and perhaps of dinosaurs. It looks as if pneumatization was a common or easily-developed diapsid feature which was seized on by birds (and pterodactyls) where it could be used in flight. Its original function remains unknown, but connections with respiration and temperature regulation have been suggested.

2.2 MUSCLES

The wing is moved by muscles, and it is obvious to anyone who has carved a fowl that it has a relatively huge muscle stretching from the sternum to the humerus. This, the pectoral (or pectoralis major), has its origin chiefly on the sternum and its keel, but also to some extent on the clavicles, and inserts on the proximal portion of the humerus; its contraction draws the wing downwards and forwards. In reptiles the corresponding muscle has also an origin from the pelvis, but this is small in birds or absent altogether, so helping to free the wing from the hindlimbs. In some birds, such as snipe, the pectoral muscles may make up one third the weight of the body; usually they are about one fifth.

Covered by the pectoral is a smaller muscle, the supracoracoid (also called the pectoralis minor, and possibly homologous with the supraspinate of man), which originates chiefly on the sternum in the angle between keel and body, but sometimes also on the procoracoid and clavicle, and in contraction pulls the humerus backwards and upwards. In order to do this it has a long tendon which runs upwards, and then turns through the gap left between clavicle, procoracoid and scapula to insert on the trochanter of the humerus from above (Fig. 2.10).

The supracoracoideus is small in many gliding birds, such as buzzards, and large in those with a quick take-off, such as pigeons and pheasants. At its largest, in humming-birds, it is about 10 per cent of the mass of the body.

The distal parts of the forelimb have little muscle, their bones being moved by tendons from muscles whose bellies are situated proximally. Small muscles of various homology fold the patagial membrane of the wing. The first digit is moved by a number of small muscles that have their origin on the radius and ulna or on the digit's own metacarpal or on the carpometacarpus.

The homologies of the muscles of the leg with those of mammals are

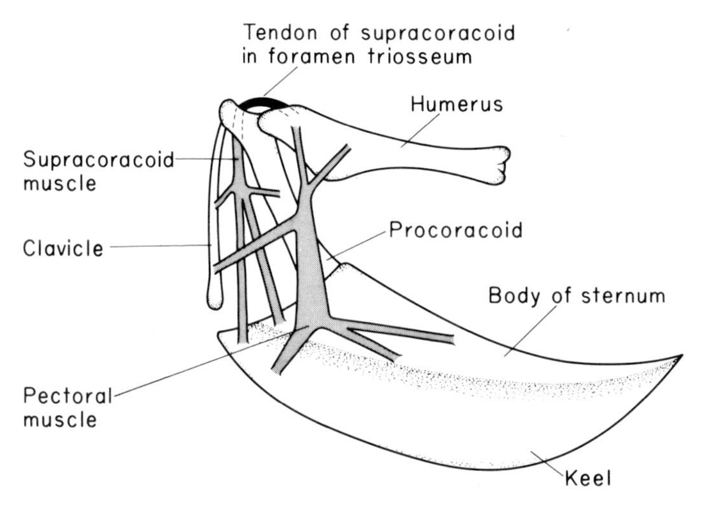

Fig. 2.10 Pectoral and supracoracoid muscles. (Based on Grassé, 1950, *Traité de Zoologie:* (15) *Oiseaux,* Fig. 86, Masson et Cie, Paris.)

doubtful. Their great characteristic is the concentration of their fibres in the proximal part of the limb, so that the tibia and more distal parts are moved by long string-like tendons. The extreme of this is seen in the ambiens (a muscle found only in crocodiles and some birds), which has its origin on the pelvic girdle and tendons stretching down to the toes (Fig. 2.11). No part of the locomotion of the bird is carried out by bending the trunk, and the muscles connecting the vertebrae, so conspicuous in a mammal, are virtually absent from the bird except in the neck.

2.3 FEATHERS

Besides a skeleton and a source of power a wing needs a flat surface to give resistance to the air. This is provided in birds to a small extent by a fold of skin, the patagium, but mainly by the feathers. They cover almost the whole of the body, and in other parts serve other purposes; those that cover the wing have what is usually regarded as the typical structure. They are named according to the bones that give them support, thus:

Bone	Feathers		
Humerus	Humerals		
Ulna	Secondaries or cubitals		remiges
Metacarpal 2	Metacarpal quills	primaries	remiges
Digit 2	Digitals	primaries	remiges
Digit 1	Bastard wing		

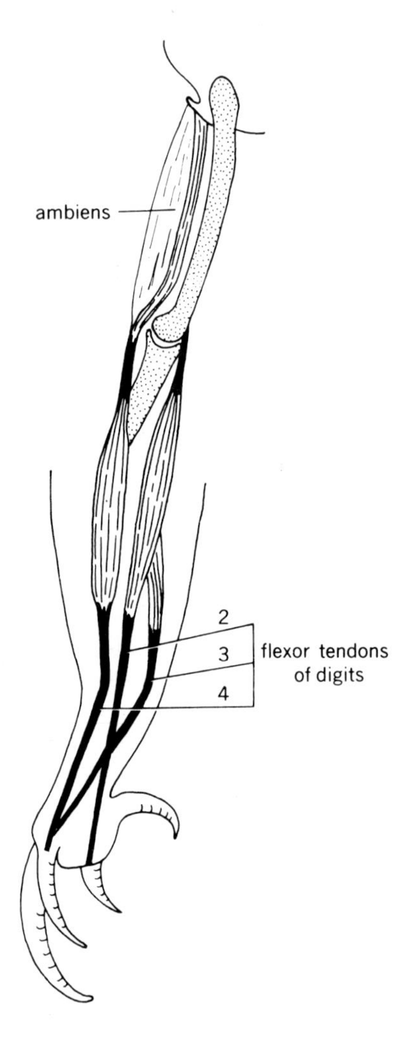

Fig. 2.11 Semi-diagrammatic drawing of a dissection of the ambiens muscle of a pigeon, *Columba livia*, ×4/3. (From W. B. Yapp, 1965, *Vertebrates: their structure and life*, Fig. 21.10, Oxford University Press, New York.)

The secondaries and primaries, which are the chief flight feathers, are firmly based in the bone; the whole wing is covered, above and below, by small feathers called coverts, which are based in the patagium.

The structure of a wing-feather is shown in Fig. 2.12. The hooks on the anterior barbules hold on to the posterior barbules and lock the barbs together so that the feather makes a firm but flexible surface for beating the air.

2.4 THE MECHANICS OF FLIGHT

Aerodynamics is a difficult science, and no complete account of the action of a bird's wing has ever been given; if it could be, few readers of this book would understand it. At its simplest, a bird's wing may be regarded as a plane moving through a fluid, the air. The forces that such movement generates are shown in Fig. 2.13, in which the plane (or wing) is shown in section. In (a) the wing is shown moving downwards, in (b) forwards as in a glide. In both cases the force of reaction of the air is, by Newton's third law, equal and opposite to that exerted by the moving wing on the air. It may be resolved in two directions at right angles, one parallel to the surface of the plane and one perpendicular to it. If the plane is thin the former will exert no effect on it and may be neglected. The second may be resolved again into horizontal and vertical components.

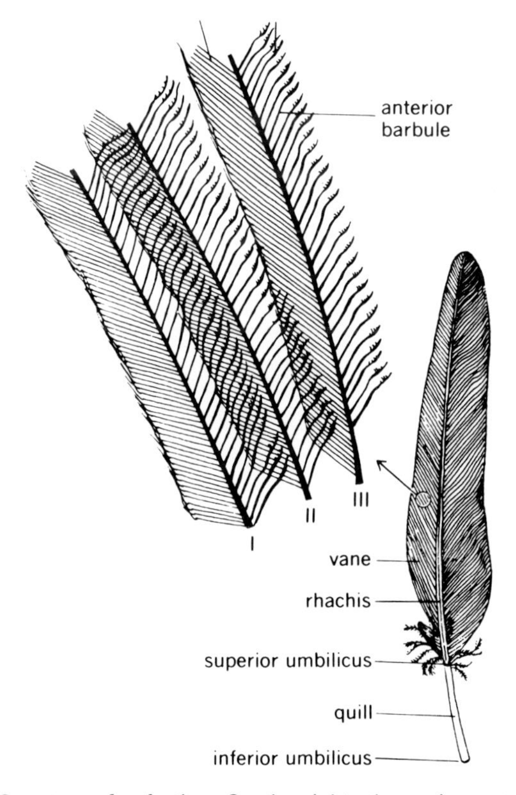

Fig. 2.12 Structure of a feather. On the right, the main parts of a typical quill; on the left, a small part of the vane, ×100. (From W. B. Yapp, 1965, *Vertebrates: their structure and life*, Fig. 18.6, Oxford University Press, New York.)

In (a) these are directed forwards and upwards, and so will propel the bird forwards and raise it (or restrain its falling, according to its weight). In (b) the vertical component is upwards, and so will again tend to lift the bird, but the horizontal component is backwards, and so will tend to slow the bird down.

Flight is complicated by two things that cannot be treated simply. First, round the edge of the wing there is turbulence, so that forces are not

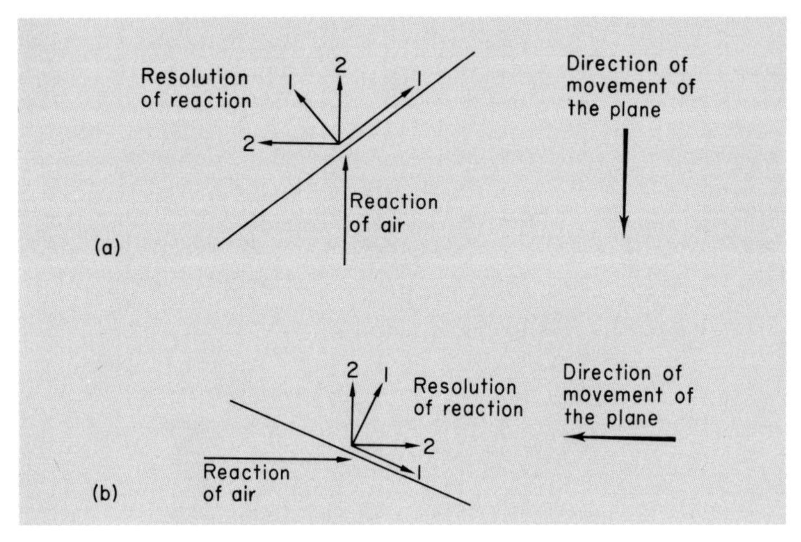

Fig. 2.13 (**a**) and (**b**) Flight diagrams.

regular in their disposition—they may, for instance, induce flutter of the wing itself. One simple statement can, however, be made about turbulence: the kinetic energy of the fluid that it involves must be got at the expense of an equal amount of kinetic energy of the wing, so that it must slow the movement down. Second, the wing is not a simple plane; it has thickness, and is generally convex above and concave below. This means that when it is moving at an angle through the air there is a reduction in air-pressure above the upper surface and an increase below, both of which will help to drive the bird forwards and upwards.

2.41 Flapping flight

All birds fly by flapping their wings, and most of them do so nearly all the time they are flying. It is easy to see that a wing moved downwards will tend to lift the bird, and one moved backwards will drive it forwards; possibly a singing lark or a rocketing pheasant, both of which fly vertically upwards, may move in the former way, but probably no bird beats its wings directly backwards. The simplest compromise is a wing-stroke directed obliquely backwards and downwards; if the bird is already moving forwards the movement of the wing relative to the air will be approximately that of Fig. 2.13a, and there will be both lift and thrust. If in the recovery stroke the wing is folded so that the area of the wing is reduced the contrary forces will be small and the net result is that the bird will be propelled forwards and sustained in the air. Such movement of the wings has been shown to occur in a few species in fast flight, and is probably general at high speeds.[51]

A peculiar variant of this is found in hovering humming birds. The body is vertical, and the wings are swept forwards (ventrally) in the position of Fig. 2.13b, and then returned so that the dorsal surfaces are facing obliquely downwards and swept backwards (dorsally). Both strokes produce lift and the forward component of the first stroke and the backward component of the second cancel each other out. Other small birds such as house-sparrows and willow-warblers, that occasionally hover, probably do so in much the same way. In humming-birds the supracoracoid muscle, which produces the backward (dorsal) stroke, is proportionately very much larger than in other birds.

Slow flight is more complicated. The down stroke produces lift but little thrust, but by a rotation of the joints there is a backward flick of the wrist and primary feathers at the end of the recovery stroke which drives the bird forwards.

2.42 Gliding and soaring

Gliding and soaring, though less common, are relatively simple. Once the bird has acquired a horizontal velocity its spread wings are in the position of Fig. 2.13b, and the forces are those of lift, preventing the bird from falling, and drag, slowing it down. By an alteration in the angle of attack (that between the plane of the wing and the direction of motion), the relative proportions of the two can be changed. Maximum lift is in practice given by an angle of attack of about 15°. Theoretically it is higher but turbulence brings it down, and anything that reduces this enables higher angles, and so lower speeds, to be used before lift becomes much less than weight and the bird loses height uncontrollably, or stalls.

If, in a glide, the lift is greater than the bird's weight, it will rise in the air, and the potential energy that it thus acquires is derived from the kinetic energy of its movement; in other words, it must slow down. If there is an upcurrent of air the bird may rise without loss of kinetic energy, and the flight changes from gliding to soaring. Such vertical currents are found where the ground is hot, when they are called thermals, and where the wind has to rise over an obstacle such as a cliff; the first type is used by vultures in hot climates, the second by the fulmar (*Fulmar glacialis*). A gliding bird can also rise without loss of speed if the air velocity relative to the ground increases with height. This may happen over the sea, and is the type of flight used by albatrosses.

2.43 The slotted wing[135]

Turbulence is reduced by smooth contours, and also by gaps in the wing, especially in the leading edge. Such gaps, called Handley Page Slots, may be seen in action in an aircraft coming in to land. In the bird the slotted effect is usually achieved by the separation of the primaries, as may be seen in many large soaring birds such as eagles and vultures. The size of the slots is increased because the anterior part of the base of some of the primaries is reduced or emarginated. Many small birds have similar emargination (Fig. 2.14) and when they are slowing down to perch the primaries are widely spread (Fig. 2.15). A third method of

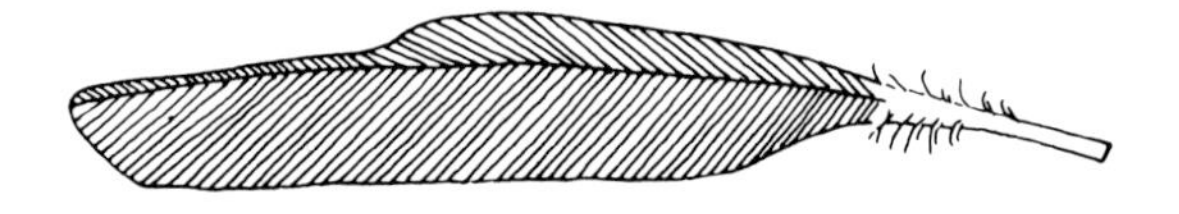

Fig. 2.14 Emarginated primary feather of a chaffinch, *Fringilla coelebs*, ×1.0.

producing a slot is by raising the bastard wing or alula, a separate tuft of feathers on the pollex. This can be seen in action when a hovering kestrel momentarily ceases to flap its wings.

In general, birds of the wide open spaces, such as waders and gulls, have little or no emargination and a small alula; they can land at high speeds and do not need slots. The lapwing (*Vanellus vanellus*), although a wader, has emarginated primaries without which it could presumably not carry out its slow acrobatic display flights. All the common British birds living amongst trees, except the cuckoo, have well-developed slots.

The ratio of the span of the wings to their breadth is the aspect ratio; it ranges from about 5 to about 20. It is clearly important, but in only a few cases is its exact importance known. A high aspect ratio means less turbulence round the ends ('induced drag') and so facilitates high gliding speeds; these are necessary for soaring by means of differential wind-speeds, and the albatross has the highest aspect ratio of all birds. But it is high also in swifts, with fast flapping flight, but low in the sparrow-hawk, also with fast flapping flight. In the last case the bird's habitat in woods would presumably make long wings impossible.

The tails plays some part in flight; it is often depressed for braking before landing.

(a)

(b)

Fig. 2.15 A swallow landing. Photographs by S. C. Porter. (**a**) shows the alula open, and (**b**) shows the primaries separated to make Handley Page Slots.

2.44 The speed of flight[208,244]

The speed of bird flight has been greatly exaggerated. Most small species have a true air speed in normal flight of about 20 mph, larger species from 30 to 40 mph, and even the fast fliers, such as swifts, falcons, ducks and pigeons, seldom exceed 50 mph.

2.5 MUSCULAR CONTRACTION

The muscular contraction by which the wings are moved is basically similar to that of mammals and other vertebrates, but birds appear to have some specialization. The nerve-muscle-junctions are mediated as usual by acetylcholine. The distinction between dark and white muscle, which depends mainly on the presence or absence of haemoglobin, is obvious in the fowl, where the white muscles of the breast and wing have little to do, since the bird hardly flies. More important is the presence in the breast muscles of many birds, from parrots to pigeons, of two types of fibre; one is white, broad and contains glycogen but no fat and few mitochondria, the other is red, of half the diameter, and contains little glycogen but much fat and many large mitochondria.[124] There is little doubt that the second type is used for sustained flight, and that fat, not carbohydrate, is the fuel. Some strong fliers, such as bee-eaters and humming-birds, have only the fat-loaded fibres, and in the latter they are present in the supracoracoid as well; in hovering, this muscle must do as much work as the pectoral. Many insects, especially those with sustained flight, such as Lepidoptera and migratory locusts, also use fat as their fuel,[390] and about two-thirds of the energy of contraction of the breast muscles of bats comes from fat.[123] The advantage of fat over carbohydrate for a flying animal is that from the same weight of material twice the energy can be produced.

Investigation of the metabolism of a flying bird is extremely difficult.[201] The few good results available suggest, as would be expected, that the output of energy is high. The pectoral muscle of a budgerigar (*Melopsittacus undulatus*) was found to be able to develop about the same power for a given weight as could that of insects, and about ten times as much as human muscle.[372]

2.6 FEATURES ASSOCIATED WITH FLIGHT

A new habit, such as was flying in the first birds, may develop more easily if certain features of anatomy or physiology are already present, or it may, in its turn, facilitate the evolution of further new features. The rest of this chapter will deal with some characteristics of birds which, while not necessary to flight, make it easier or more successful, and so are likely to be linked with it in evolution either as cause or as effect or as both.

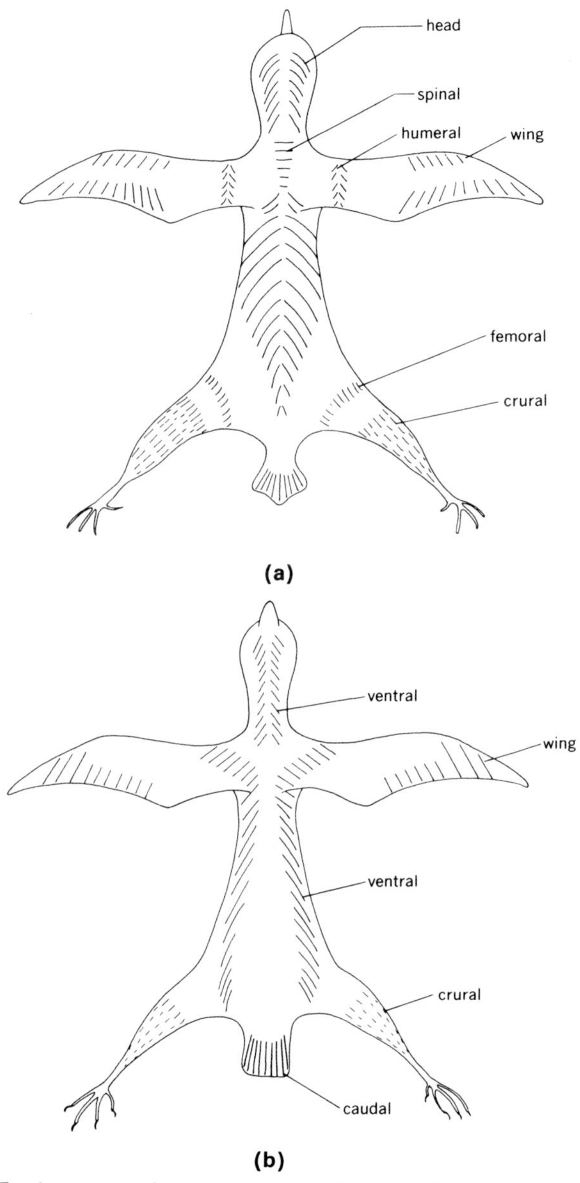

Fig. 2.16 Feather tracts of a young thrush, *Turdus philomelos*, diagrammatic, (**a**) dorsal (**b**) ventral. The names of the tracts are given. (From W. B. Yapp, 1965, *Vertebrates: their structure and life*, Figures 18.11 and 18.12, Oxford University Press, New York.)

2.61 Temperature control

Flying needs a high output of power, and the chemical reactions on which this is based will go on more rapidly the higher the temperature; equally, they will themselves produce waste heat, which will raise the temperature of the body. Either way, a flying animal is likely to have a temperature well above that of its surroundings, and it is likely to benefit (as does a motor car) if it has a thermostat, or, in biological language, if it is homoiothermic. Birds share with mammals an ability to maintain their temperature constant within quite narrow limits, and although this temperature control must have been acquired independently by the two groups the methods by which it is achieved are remarkably similar. Further reference is made to this in Chapter 4.

We do not know if the non-flying ancestors of the birds had temperature control, but there are good reasons for thinking that some of the dinosaurs, and perhaps the pterodactyls, were warm-blooded. If this were so, it looks as if there was a tendency for this to occur in the diapsids, and it becomes more likely that temperature control developed first, and was a pre-condition of flight, than that the excessive heat production of flight imposed temperature control. This view is reinforced by the presence of feathers on the general body-surface of *Archaeopteryx*. They could have had no function other than that which they have on the body of modern birds: to act as an insulating layer. Such a layer does not necessitate homoiothermy, but an insulating layer and temperature control are much more likely to evolve together than is either of them by itself. It is therefore at least possible, if not probable, that the immediate ancestors of birds were warm-blooded, and that the primary function of feathers is not flight but insulation.

If this is so, the typical feather is not the large quill of the wing, but a smaller type such as one of the contour feathers that cover the body. These grow in patches, called pterylae, separated by areas, apteria, that bear only down (Fig. 2.16). In some species and situations insulation is improved by the presence of specially modified feathers, the down, beneath the contour feathers (Fig. 2.17). The total number of body feathers in a small bird such as a sparrow is about 3000.

2.62 Colour

In many birds feathers have taken on a third function of bearing colour, which may have value or meaning in several different ways. Where the sexes are different, as in the bullfinch (*Pyrrhula pyrrhula*) and even more markedly in birds such as pheasants and most of the weaver-birds, one may assume that there is some value in sexual selection. The red breast of the robin (*Erithacus rubecula*) is the feature by which other robins recognise one of their own species; the speckled brown of a hen partridge

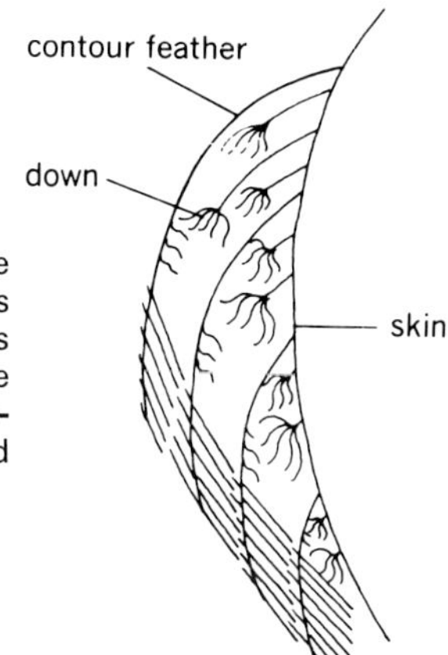

Fig. 2.17 Diagrammatic section through the breast feathers of a swan. The contour feathers have a normal vane distally and plumaceous unhooked barbs proximally, so trapping a large amount of air. (From W. B. Yapp, 1965, *Vertebrates: their structure and life*, Fig. 18.9 c, Oxford University Press, New York.)

(*Perdix perdix*) or the white, black and brown of a ptarmigan (*Lagopus mutus*) amongst the rocks and snow can hardly be doubted to be cryptic. The white flash of the tail of a wheatear (*Oenanthe oenanthe*), or the similar white shown by many waders when they fly, is perhaps a distraction signal, causing confusion in a predator when the bird suddenly settles and is no longer visible.

Many of the colours of birds are due to pigments, such as melanins, carotenoids and porphyrins, chemically related to those found in other animals, but some of the most brilliant colours, especially blues, are caused by the minute structure of the feathers.[19,114] The colours of humming-birds, pigeons, peacocks and glossy starlings are produced by interference at the surface of granules of melanin. The blues of kingfishers and blue tits (*Parus caeruleus*) are caused by dispersion (or Tyndall scattering) and so do not change with the angle of incidence in the way in which interference colours do.

2.63 Moult

Feathers are epidermal structures, and, like hairs, may be regarded as derived from the reptilian scale. Their chief constituent is keratin, the material of which the hard outer part of our skin is made. It is a protein, and as such must be derived from valuable materials that could otherwise be used differently. In most small birds the feathers make up about 10 per cent of the weight of the body,[375] and such a large amount of protein is not likely to be locked up in this way unless it is performing a useful function. Once in the form of keratin the protein is lost, and feathers are shed

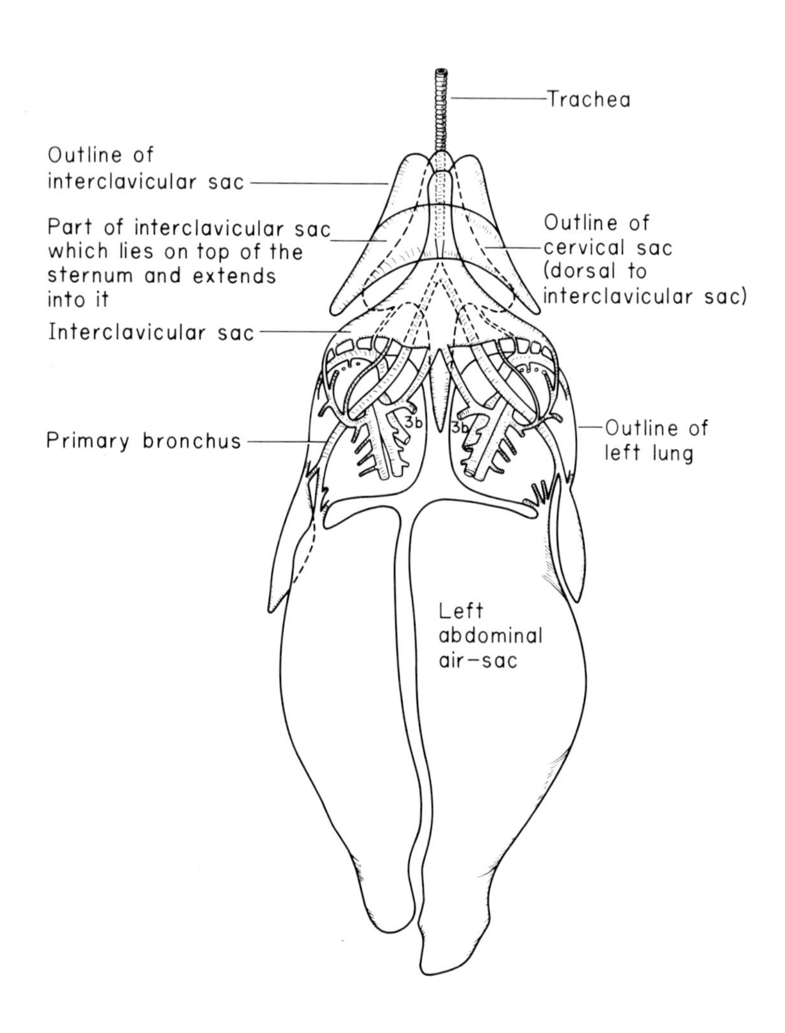

Fig. 2.18 Connections between lungs and air-sacs in the fowl (ventral view). Both anterior thoracic sacs have been removed. The row of four small holes in the anteriolateral corner of each lung represent tertiary bronchial connections with the anterior thoracic air-sac. The large duct (3*b*) from no. 3 anterior dorsal secondary bronchus also connects with the anterior air-sac. (After A. R. Akester, (1960), Fig. 2, *J. Anat.* **94**, 487.)

annually (a process called moult) and replaced. Some feathers are lost throughout the year, but there is also one season, or sometimes two, when almost all the feathers are lost and replaced during a period of a few weeks. When all the flight feathers go together, as in many ducks, the bird cannot fly and is likely to be subject to high predation. More often, a few feathers only are lost at a time. Presumably the primary function of moult is the replacement of dead structures, which are necessarily subject to wear.

2.64 Respiratory system

The intense activity of flight requires large amounts of oxygen, and the respiratory system of birds is more complicated than that of any other vertebrates. The trachea divides as usual into a bronchus for each lung; in the bird these are called primary bronchi. Each primary bronchus divides again into many secondary bronchi, which may be distinguished as dorsal, ventral and lateral according to their position. The secondary bronchi divide again into many tertiary bronchi or parabronchi, which have fine branches or air-capillaries with very thin walls. So described, the lung does not sound much unlike that of a mammal, but it differs in one important feature: at all levels from the secondary bronchi on there is a greater or lesser degree of anastomosis. Some of the secondary bronchi fuse; all the parabronchi join with their neighbours so that none ends blindly; and the air-capillaries anastomose, especially in birds of strong flight such as pigeons and buzzards. The result of this is that the dead space, through which air does not flow, is very small or non-existent.

The relative volume of a bird's lung is small, less than half of that of mammals. In some way this is presumably compensated for by the presence of air-sacs, which are non-muscular and non-vascular extensions of the lungs[15] (Fig. 2.18). There are slight differences between species, but in all the general pattern is the same. The cervical, interclavicular, anterior thoracic (or thoracic), and posterior thoracic (or anterior abdominal) sacs are expansions of secondary bronchi in appropriate positions, while the abdominal (or posterior abdominal) sac arises at the posterior end of the primary bronchus. All are paired in origin, but in the pigeon the two interclavicular sacs fuse early in development and in the fowl the cervical sacs fuse also. From the air-sacs, especially the cervical and interclavicular, are given off the extensions that have already been described as invading the bones; the exact pattern depends on the species. Most of the sacs have only a single connection to the bronchi, but a few have subsidiary connections.

2.641 Breathing

In spite of the great interest of the subject, we do not yet know how the system works. It is obvious that in breathing the sternum moves up and down, and in doing so it greatly alters the volume of the air-sacs while having little effect on that of the relatively firm lungs. Flapping the wings

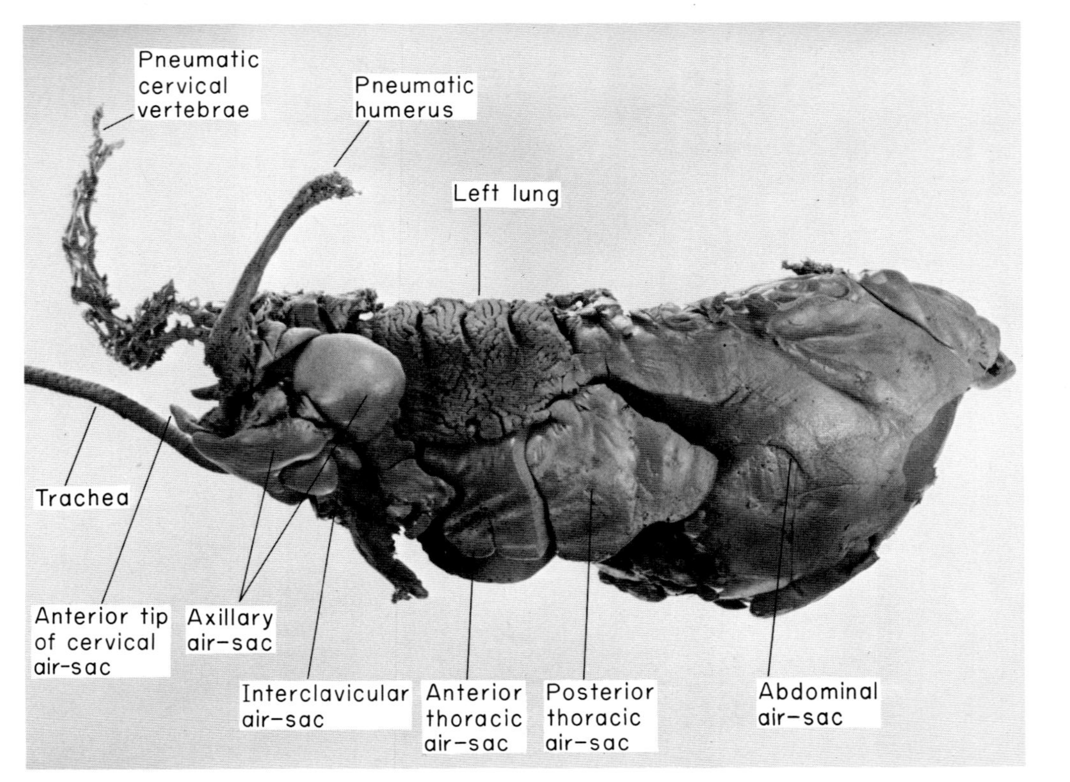

Fig. 2.19 Whole cast of lung and air-sac of fowl, injected with rubber latex; left lateral view. (From A. R. Akester, (1960), Plate I, *J. Anat.* **94**, 487.)

must also alter the volume of the body cavity, and in the pigeon breathing is synchronized with the wing-beats. What is not certain is the course of the air within the system. Some experiments have been claimed to demonstrate that there is a circulation of air through the lungs; through the primary bronchus to the abdominal and thoracic sacs and then out by the subsidiary connections ('recurrent bronchi') and parabronchi, but this cannot apply to the majority of the sacs. According to other authors the sacs draw air through the parabronchi on inspiration, and, because of the aerodynamic layout of the system, blow it through them in the same direction on expiration.[156] In any case the main flow seems to be from the dorsal secondary bronchi through the parabronchi to the ventral secondaries, and whether it is continuous or intermittent the ventilation is much better than in a mammal. The total volume of the sacs is several times that of the lungs.[184] A second function of the sacs is probably to allow for the movement of the heart in an otherwise very rigid thorax.[264] The air-capillaries, the actual site of gas exchange, are intermixed with blood capillaries.

2.65 The senses

The faster an animal moves, the more necessary it is for it to receive accurate and rapid information about its surroundings. All birds, if they fly at all, can achieve speeds comparable with those achieved on the ground only by the fastest of mammals; moreover, a collision in the air is likely to be much more harmful than one on the ground. It is therefore to be expected that birds should have evolved good distance receptors and a good sense of balance. In doing so, they have used the structures already in existence in reptiles and have developed them to an extent rarely or never reached in other groups.

2.651 Sight

The chief distant sense is sight. Man is himself a visual animal, and is apt to forget how poorly most animals are equipped in this respect. Birds alone have better eyes than he has. They are, in general, diurnal, and use their eyes for finding food, recognizing their mates and so on. Though these uses have no doubt affected the evolution of the eyes in some respects, the necessity of good vision remains primarily connected with flight; without it a bird could neither avoid obstacles nor know when and where to land, so that flight would be impossible. The only possible alternative that we know of is the sonar used by bats and, indeed, by a few species of bird. The details of vision in birds are discussed in Chapter 4.

2.652 Hearing

The other distant sense on which man relies, although to a much less extent than on sight, is hearing. Birds have good hearing, (see Chapter 4), but only in a few species has it any very obvious connection with flight.

Birds have twice developed echolocation. The oilbird (*Steatornis capensis*) of Trinidad and South America, and some swifts of the genus *Collocalia* of south-east Asia, and Australia, all nest in dark caves, and find their way by giving out short clicks and responding in flight to the reflection of these from the walls.[137] The oilbird's clicks have a mean frequency of about 7000 Hz, and those of the swift are about an octave lower. There seems to be nothing in birds like the echolocation by ultrasonic sounds of the order of 100 kHz used by bats for finding the insects on which they feed. For reflection from the walls of a cave longer waves are good enough. The short length of the bird's cochlea suggests that ultrasonic hearing would be difficult, while it occurs in several mammals, such as rodents, shrews and whales, as well as in bats.

2.653 *Balance*

The parts of the inner ear connected not with hearing but with balance do not differ greatly from those of other vertebrates, and are presumably used in the same way. Maintenance of the correct posture and rapid correction of deviations must be even more important in flight than in walking, and there can be no doubt that birds use their ears in this way. They will be assisted by their eyes, as in man; in the days before automatic control an aircraft in mist might be flown tilted to right or left or nose up without the pilot knowing. The unwillingness of birds to fly in mist suggests that in them too the eyes may be of great importance in balance, and we shall see in section 4.44 that the horizontal fovea may be used to keep the bird on a level course.[208]

2.66 The brain

Good sense organs demand a good central nervous system. Intelligence, however it may be defined, needs more than a large quantity of information, but it cannot exist without it. The input from the sense organs, and the means of processing it, increase in evolution hand in hand.

The brain of birds is roughly ten times the size of that of reptiles of the same size and comparable with that of mammals. The birds with the relatively largest brains are the hawks, owls, woodpeckers, song-birds (especially the crows) and the parrots, and the smallest are found in the game-birds and pigeons and (if the European species is typical) the nightjars.[199,290] The increase in size and differences from the reptiles are mainly in three parts of the brain (Fig. 2.20). The cerebral hemispheres are greatly expanded, both in their ventral or striatal portion and in the dorsal cortex. In the latter much of the expansion, as in mammals, consists of a development of the neopallium, which is scarcely represented in reptiles. Both in birds and mammals the cortex is largely concerned with

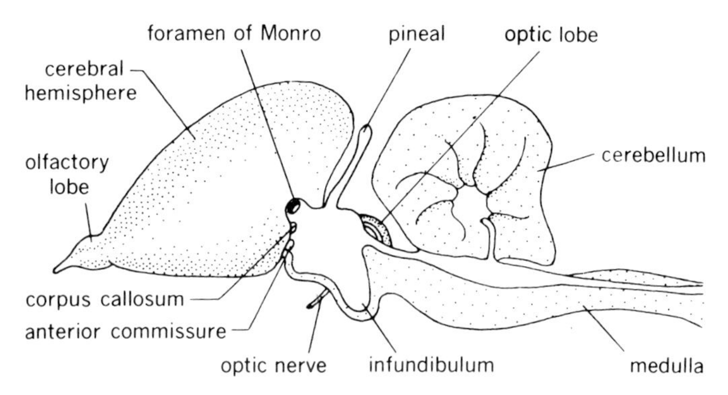

Fig. 2.20 Diagram of the right half of the brain of a pigeon, *Columba livia*, cut sagitally, ×2. The cerebellum has been displaced slightly backward to expose the optic lobe, which it covers in the natural position. (From W. B. Yapp, 1965, *Vertebrates: their structure and life*, Fig. 13.14, Oxford University Press, New York.)

the formation of conditioned reflexes. Pigeons from which the hemispheres have been removed can still see and fly, but they cannot form conditioned reflexes and have lost those they formerly possessed. The part of the cortex called the palaeopallium, which in mammals is concerned with smell, is small, as generally are the olfactory lobes.

The optic lobes are large, and are pushed to the sides by the constriction of the orbits and the increased size of the hemispheres. Most of the fibres of the optic nerves end in them, only a few going forward to the cerebral cortex, so that there is little that corresponds to the visual cortex of mammals. Presumably instead of this they have a more complicated internal structure than do those of mammals.

The cerebellum is relatively larger even than in mammals, and has some surface fissures and considerable complication of its cellular structure. It is concerned largely with movement and balance. The pterodactyls (in which, as in birds, the inside of the cranium bears impressions that can only be interpreted as those of the brain fitting tightly within it) also had a large cerebellum, reaching forward over the mid-brain to reach the cerebrum.

To sum up this Chapter, we may say that birds were able to evolve because their reptilian ancestors had certain properties, both of anatomy and physiology, that made flying relatively easy. Flight meant that some modifications, especially in the sense-organs, were particularly valuable, and so were selected, and these modifications themselves increased the value of other features, such as an enlarged brain. Characters from all four levels—feathers, flight, good eyes and ears, an enlarged brain—co-operate to make possible that higher life of complex behaviour which makes birds so different from reptiles and which is discussed in Chapter 6.

3

Classification and Adaptive Radiation

One of the biggest difficulties in the classification of birds is that, with the exception of *Archaeopteryx*, their structure is very uniform. There are some 9000 species of living birds, but they show nothing like the range of form shown by the mammals, which have only half that number. Indeed, the birds show hardly more differences than can be seen within the Primates, with about 90 species, and certainly no more than there are within the 290 species of the Carnivora, if one includes the seals in that order. To be fair, we must add that we are remarkably ignorant of the comparative anatomy of birds, the study of which has not kept pace with the great increase in our knowledge of habits and behaviour. Macdonald wrote in 1959[223] that the structure of the pulmonary system was known for fewer than 12 species, and anyone who wants to read about the internal anatomy of birds has to go to books written seventy years ago.

Birds are classified, on present knowledge, chiefly by means of obvious external form and some features of the skeletal and muscular systems. In this way living birds are divided in the most recent classifications into some 27 orders, with two others recently extinct.[394] This is nearly twice the number recognized at the end of the last century. These 'orders' are about the equivalent in rank of mammalian families, and their number is a reflection of the number of bird species (and perhaps of bird taxonomists) rather than of any great distinction in structure between their members.

The orders of birds are, with few exceptions, generally accepted, but their association into larger groups (superorders or subclasses) is still controversial. It was for a long time customary to associate many flightless birds together as the subclass Ratitae and the others as Carinatae, but the form of the foot and other features are different in the main species of the former, and many zoologists now think that flight has been lost several times, and that the ostriches, rheas, emus and so on are not closely related. Even if they are, their differences from the others probably do not justify more than superordinal separation. More recently some taxonomists have separated the penguins as a subclass (or at least a superorder), the Impennes. Certainly their wing skeleton differs somewhat from that of a normal bird, but much less, for example, than does that of the limb of a bat or whale from the usual mammalian pattern. This leaves us only with the subclass or superorder Odontognathae, erected to include the extinct toothed birds *Hesperornis* and *Ichthyornis* and their relations. These two are not closely related, and if they did not in fact possess teeth there seems no reason to separate them, except ordinally, from modern birds. The class Aves may then be divided into the subclass Saurornithes (= Archaeornithes), containing only *Archaeopteryx*, and the subclass Neornithes, divided into about 30 orders and containing all other birds. Most taxonomists now agree in naming the orders by adding the termination '-formes' to the root of a characteristic generic name. Family names are similarly derived, in the standard zoological manner, with the termination '-idae'.

Attempts have been made in the past to group the orders according to the presence or absence of some particular feature. For example, those with an ambiens muscle in the leg were called Homolognatae, those without, Anomolognatae;[120] the carinate birds were placed in four groups (Dromaeognathae, Schizognathae, Aegithognathae, Desmognathae) according to the form of the palate. All such simple groupings lead to some absurd conclusions, or have to admit exceptions, and have long been abandoned. They are mentioned here only because they linger on, and are made to appear important, in some semi-popular books on birds. Taxonomists try to take account of as many characters as possible, and in the end have to use their own judgement; even if they believe in 'numerical taxonomy', in which a sort of balance-sheet of characters is made, they still have to judge what is a character, before the enumeration can begin.

Zoologists generally agree that characters that are easily altered by natural selection are of less taxonomic importance than others; that is, differences in them will justify lower rankings, generic or familiar, while ordinal separation must be based on less easily altered, and often internal, features. While this principle is probably sound, it is often difficult to apply, for in some features birds are extraordinarily plastic. Blood vessels, for instance, would seem to be parts of the body on which natural selection would act only slowly, and in mammals this expectation is fulfilled. In birds one or other of the carotid arteries is often suppressed, and although the pattern generally follows the classification arrived at on other grounds, there are exceptions for which no explanation can be given.

3.1 MODERN TAXONOMY

Classical taxonomy is based on a study of morphological characters, but even in the nineteenth century other types of character were sometimes used, and in recent years three have been especially stressed.

Nearly all animals have parasites, many of which are restricted to one species or one genus of host. Where this is so, it follows that as the host-group has evolved and split into new genera or species, the parasites must have differentiated in the same way. If this argument is put into reverse, one may say that where the relationships between parasites are clear they may be used to determine relationships between the hosts.[64] The tapeworms of penguins, for example, are similar to those of petrels and pelicans, and so Baer has suggested that the three groups are related.[21] On the other hand, all three are marine and fish-eating, so that the worms may have adapted to similar ecological conditions in the guts of their hosts. The lice (Mallophaga) suggest relationship between the gulls, waders and auks, which were first united in the nineteenth century, on morphological and other grounds, as the order Charadriiformes.[65]

No one doubts that morphological characters are largely based on genes, and therefore fundamentally biochemical, so that from time to time attempts have been made to use biochemical characters, mainly of proteins, as classificatory data. The immune reactions of serum have been used in this way, not always successfully,[410] and more recently the patterns produced by the movement of proteins in an electric field (electrophoresis) have been used.[320] Sibley has applied this method to the egg-white of many species, and has confirmed that the waders, gulls and auks are closely related. On the other hand he finds relationships between the owls and nightjars, which are normally placed in separate orders although they were united by Gadow and others seventy years ago, and he does not support the association of penguins and petrels (Fig. 3.1).

Both these methods may be useful adjuncts to the usual ones, but it does not seem that either is the final solution that its practitioners sometimes claim. One of the attractions of the analysis of egg-white proteins is that they are relatively simple, but this is one of their dangers. While the electrophoretic patterns for egg-white remain constant almost throughout the development of a hen's egg, those for the serum proteins change rapidly during development.[304] Similarity and difference between species may therefore represent differences in rates of development rather than anything else.

Some of the complicated patterns of behaviour have been used in taxonomy, especially in deciding the relationships of genera and families. The house-sparrow (*Passer domesticus*), for example, unlike the finches and buntings with which it was for long associated, makes a large untidy nest,

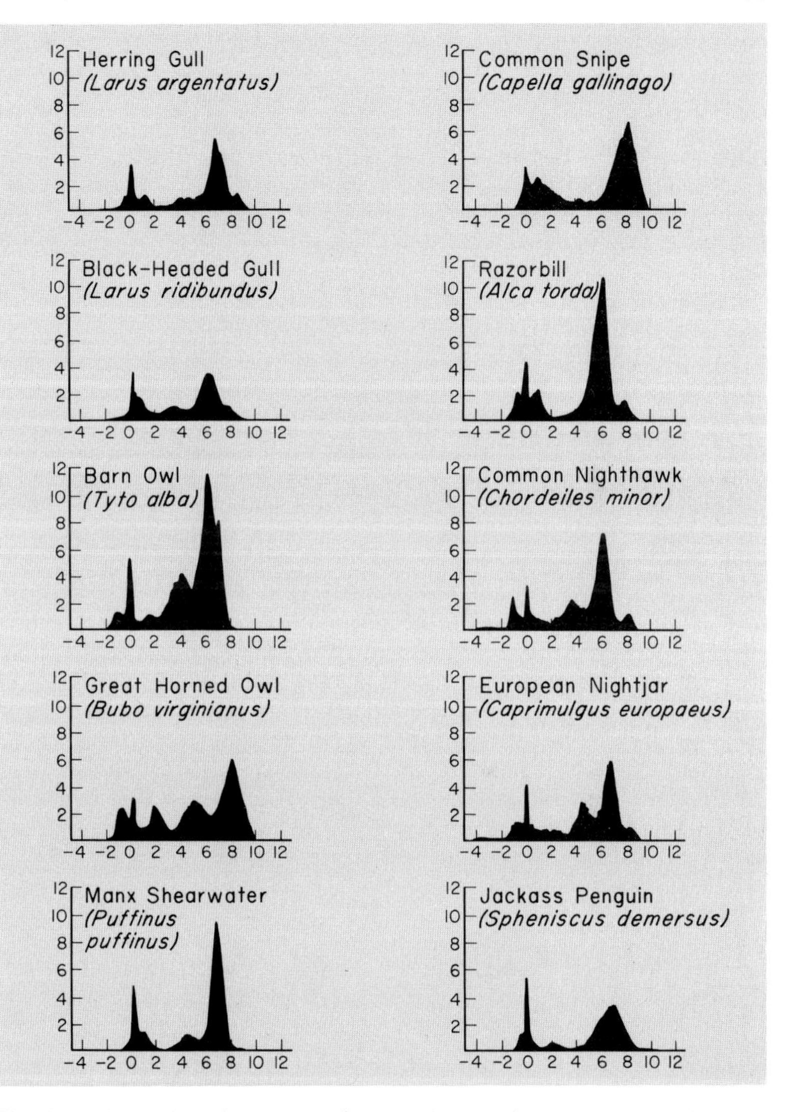

Fig. 3.1 Electrophoresis patterns of egg-white proteins. The gulls (Lari) snipe, (Charadrii), and razorbill (Alcae) are very similar, and are united in the order Charadriiformes. The owls (Strigiformes) and nightjars (Caprimulgiformes) are very similar; formerly united, they are now generally placed in separate orders. The shearwater (Procellariiformes) and penguin (Sphenisciformes) are superficially similar, but some of the troughs and peaks are in different places. (From Sibley[320], 1960, *Ibis* **102**, 215.)

of straw or grass, and this is an additional reason for transferring it to the family of weaver-birds (Ploceidae), with which it was associated some years ago on anatomical grounds.[351] But behaviour must be used with great care; it is seldom or never helpful in dealing with the larger groups, and must always give way to morphology when the two conflict. Thus the sand-grouse (Pteroclididae) resemble the game-birds in nesting on the ground, and in their cryptic eggs and active young, but are close to the pigeons in their skeleton and feather pattern; they are therefore included with the latter in the order Columbiformes.[5]

Ornithologists have spent much time in discussing the order in which the groups of birds, and especially the families within the Passeriformes, should be placed. In so doing, some of them have forgotten the purposes of classification, which are: firstly, to arrange multitudinous species so that economical statements can be made about groups of them and so that they can be readily found in museum cabinets; and secondly, to display evolutionary relationships in so far as that may be possible. The second cannot be done in a list that must run in linear or one-dimensional form, since, even though we do not know exactly how the families of birds were derived, we can be certain that their family tree has at least two dimensions, and possibly more. For the first, nothing matters except that whatever system is chosen should be consistent and easily used. There is no one best or final order, and the argument used by Wetmore, to end classificatory lists with the Fringillidae, because this group is the modern expression of a main core or stem that throughout the Tertiary periods has given rise to more specialized assemblages, seems to be the same as that which leads most mammalogists to place the Insectivora at the beginning of their lists. There is now a growing feeling that the reader is often best served by placing lists in alphabetical order, as has long been done by botanists, and as Newton did for one group of bird taxa in 1896.

3.2 ADAPTIVE RADIATION

Whatever the relationships of the orders, the members of each usually have a characteristic mode of life to which their structure is more or less obviously fitted. If we assume a single ancestry for all the birds, the evolution of these diverse ways of living and of the alterations in structure appropriate to them is called adaptive radiation. It need not, however, be a single radiation, for branching may have taken place at more than one point, and so similar habits, and up to a point similar structures, may have been evolved at different times in different groups. Just as the vertebrates have three times given rise to flying animals (pterodactyls, birds and bats), so the birds have several times over produced swimming forms and flightless forms. We shall now consider the main types of adaptive radiation, with examples, and in so doing shall illustrate some of the problems of taxonomy.

3.21 Flightless birds

Birds are characteristically flying animals, so that, paradoxically, adaptive radiation is well shown by those that do not fly. It is obvious that some birds fly better or more than others, and in general those that fly little do not appear to fly well, although some birds that hardly ever leave the ground for most of the year perform long migrations. The rails (Rallidae, including coots and moorhens) spend most of their time on the ground or in the water; unlike most birds they do not, when disturbed, fly almost at once, but walk or run away, and only rise into the air if they cannot reach shelter or water; their flight is laboured and heavy, and it is easy to suggest that it is in process of being lost. In some species it has been; most of these live on

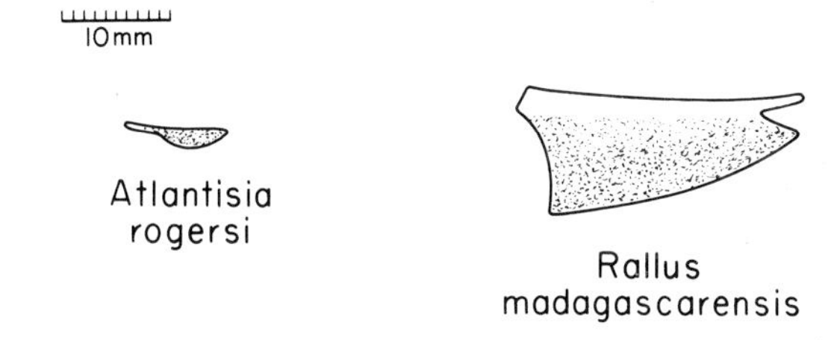

Fig. 3.2 Sterna of a flightless rail, *Atlantisia rogersi*, and a flying rail, *Rallus madagascarensis*. Length of skulls, from occiput to frontonasal suture are: *Atlantisia* 23 mm, *Rallus* 28 mm.

oceanic islands, and possibly ceased to fly because, as Darwin suggested for the comparable case of insects in 1859, in such circumstances it was an advantage to have no risk of being blown out to sea. Natural selection could only work in this case if there were no countervailing disadvantage, and in particular if there were no predators to increase the death-rate of flightless birds. This condition is in general fulfilled, but the coming of civilized man, or the rats that he has introduced, has exterminated many of the flightless rails, and reduced others to very low numbers.

In all rails the wings are small, and in the flightless forms the secondaries are missing and the keel is much reduced (Fig. 3.2).

A few members of swimming groups are also unable to fly. Two species of South American steamer-ducks (*Tachyeres*), the grebe *Centropelma micropterum* of Lake Titicaca, the cormorant *Nannopterum harrisi* of the Galapagos Islands, and the extinct great auk (*Pinguinus impennis*) of the North Atlantic are examples. All of these have relatively short wings but do not otherwise differ much from other members of the families to which they belong. The flightless penguins are referred to below.

Of the families of birds whose members live a normal life in trees, only one, the New Zealand parrot or kakapo, is known to be flightless.

3.211 *The Ratitae*

In 1813 Merrem separated a small number of flightless birds—what are now called ostriches (*Struthio*), emus (*Dromaius*), cassowaries (*Casuarius*) and rheas (*Rhea* and *Pterocnemia*)—from all others on account of their agreeing in the absence of a keel on the breastbone and in several other characters. The normal birds, with a keel, he called Carinatae, and those without Ratitae.[248] We can now add to the latter the kiwi (*Apteryx*) and the extinct moa (*Dinornis*) and elephant-bird (*Aepyornis*). They agree, with

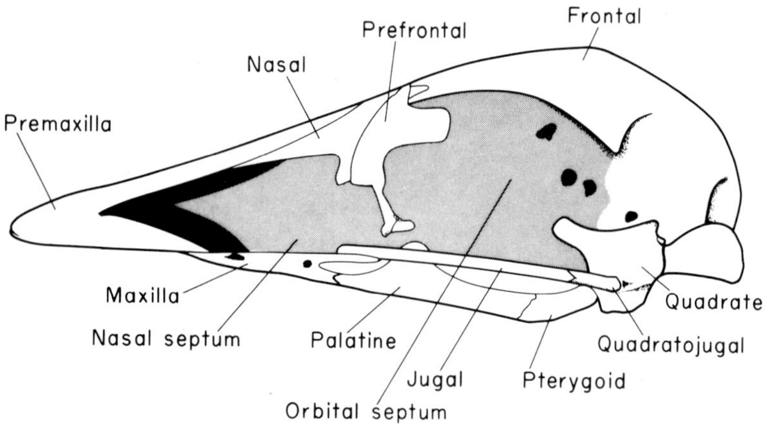

Fig. 3.3 The skull of an ostrich, *Struthio camelus*, ×1/2.

few exceptions, in a number of points:[5,38] the feathers are without hooks on the barbs and the pterylosis is similar; there is a pair of caeca with a single opening to the intestine; the first toe is absent; there is no wish-bone; the coracoids and scapulae are fused; the palate is of a peculiar type called dromaeognathous (or palaeognathous), in which the vomers (called by some authors prevomers) are long and extend from the premaxillae to the pterygoids; the orbit and external naris are confluent and the orbital and nasal septa continuous; the nasal does not reach the maxilla, and there is a peculiar form of kinetism in which the upper jaw bends upwards, not at its base, but halfway along its length (Fig. 3.3). Most of these features are found in no other birds, and the others only sporadically; several species of woodpeckers, for example, such as the three-toed woodpecker

(*Picoides tridactylus*), which is found in coniferous forests throughout the holarctic, have lost the first toe, but do not otherwise differ more than generically from other woodpeckers.

In spite of these and other smaller points of agreement all recent classifications of birds have abandoned the group Ratitae and distributed the members amongst six orders which are said to be of no specially close relationship. The only reason given for this in the current books depends on a view of the structure of the palate which has been controverted by other workers, and the similarities listed above are ignored. Such a degree of convergence seems unlikely, and the anatomical evidence strongly suggests that the Ratitae form a natural group. This conclusion is reinforced by some collateral evidence: the ostrich of Africa and the rheas of America share a genus of lice that is found on no other birds;[65] electrophoresis of the egg-white proteins suggests a relationship between ostrich, emu and cassowary, and possibly the rheas;[320] and there are some remarkable and unusual features in the birds' behaviour, such as incubation and care of the young predominantly by the cock bird.[246]

Some workers who have held that the ratites are monophyletic have believed also that they evolved from a dinosaur-like reptile independently of the carinates.[166,220] There are indeed some similarities between the skeleton of an ostrich and that of a pseudosuchian or dinosaur, but probably no more than would be expected in animals of generally similar bipedal form and habit. On the other hand the structure of the forelimb of the ratites, the arrangement of their feathers, and their large cerebellum, can hardly be explained except as being derived from those of flying ancestors.[81] Rheas, for example, have a bastard wing, which has no known function except in flight.

The characteristic common features of the ratites probably arose in a group of birds which, possibly as early as the Eocene, had already become mainly terrestrial, as have the present-day rails. They then diverged along half a dozen lines, most of which ended in flightlessness, and the differences between known forms are no more than would be expected if their final evolution largely took place, as apparently it has, in different continents; (the closely-related emu and cassowary are both found in Australia, and the two rheas in South America). One of the most striking differences is the presence of two toes only in the ostrich, and three in all the others. All agree in a considerable reduction in the bones of the wing, but the details differ; moas had no wings at all. Many muscles of the forelimb are lost, and the list is much the same for all except the kiwi, which is of different habits from the rest (Fig. 3.4).

Also except for the kiwi the ratites are large, or even gigantic, some species of moa and elephant-bird having been 10 feet tall and much heavier in proportion than any existing bird. They are mostly grazing and browsing

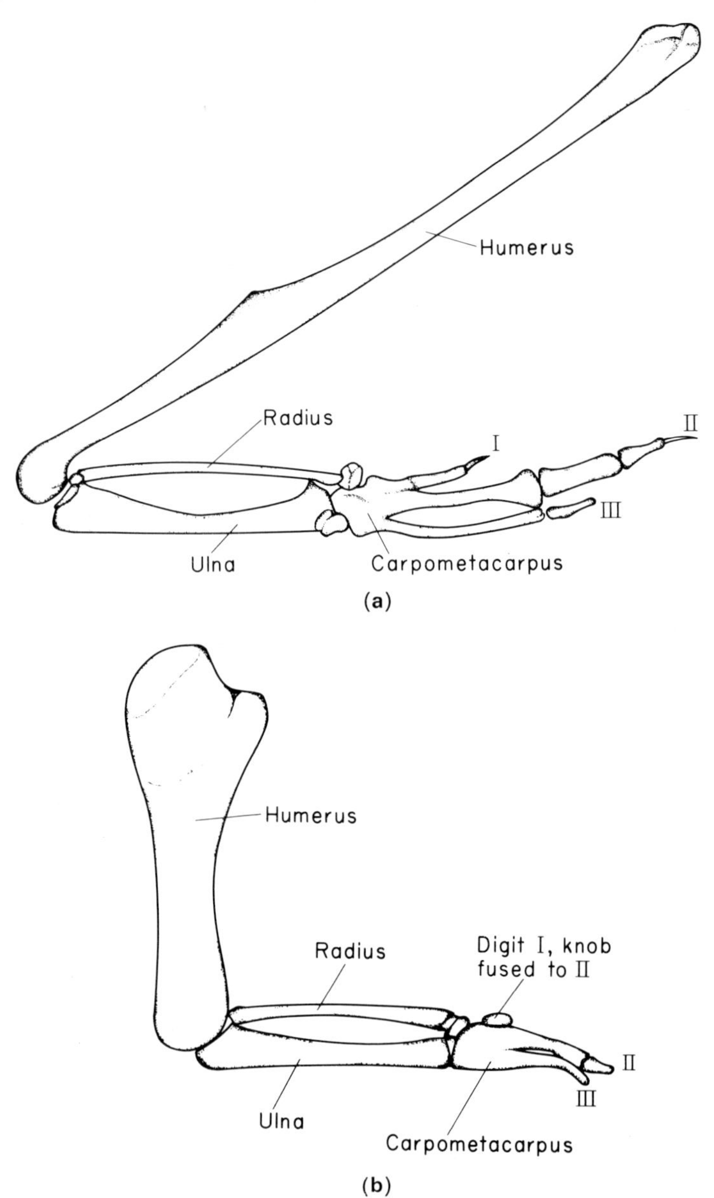

Fig. 3.4 (**a**) The skeleton of the left wing of an ostrich, *Struthio camelus*, ×1/3, seen from behind and below. Humerus=360 mm. (**b**) The skeleton of the left wing of a cassowary, *Casuarius* sp., ×3/4, seen from below. Humerus=75 mm.

animals, and fill a similar ecological niche to the large herbivorous mammals; cassowaries are more or less omnivorous. Ostriches are the only flightless birds that have been able to withstand large carnivores, and along with ungulates maintain themselves in Africa in spite of the presence of lions. In the same way the others do not seem to suffer from the smaller carnivorous mammals present in America or Australia.

The kiwis differ from the rest in being smaller, about the size of a fowl, and in feeding chiefly on small invertebrates. They have large olfactory lobes and a good sense of smell. Not surprisingly many of the details of their structure are different from those of the larger ratites; for example they have a small first toe that does not touch the ground.

It looks as if the stock from which the ratites came has left one small group of flying descendants, the tinamous (Tinamidae), a small family of ground-living birds in South America. They have a keel and can fly, but are described as crashing into branches and sometimes killing themselves in so doing.[2] They form a parallel to the evolution of the other groups, and are probably best included in the Ratitae.

3.22 Swimming and diving birds

No higher vertebrate has become fully aquatic in the sense of being able, like the larvae of many secondarily aquatic insects, to live under water without ever coming to the surface, but all three classes (reptiles, birds and mammals) have produced groups that spend most of their time in the water. The whales have their limbs so much reduced that they cannot move on land and spend all their life, from birth to death, in the sea; the same must have been true of the ichthyosaurs. No bird has gone as far as this, possibly because no bird has been able to develop viviparity; it is obvious that a large shelled egg would be at a disadvantage in water, where adequate temperature for its development would be difficult or impossible to attain (we know that the ichthyosaurs were viviparous from fossil skeletons of young found within those of adults). Ichthyosaurs and whales agree also in having returned to a fish-like mode of swimming, in which a tail-fin is moved by muscles of the trunk, but having lost the somites of the fish the whales have had to use different muscles and move the tail up and down instead of from side to side. No such mode of swimming has developed in birds, presumably because their rigid thorax makes such bending impossible.

Up to, but excluding, the step of producing their young in the water, there are all degrees of aquatic life amongst birds.[349] Many waders may get their food in shallow water, but are equally at home on dry land; some can swim, but seldom do so. The typical kingfishers plunge into the water for their food, but do not swim; they have merely adapted the pouncing habit

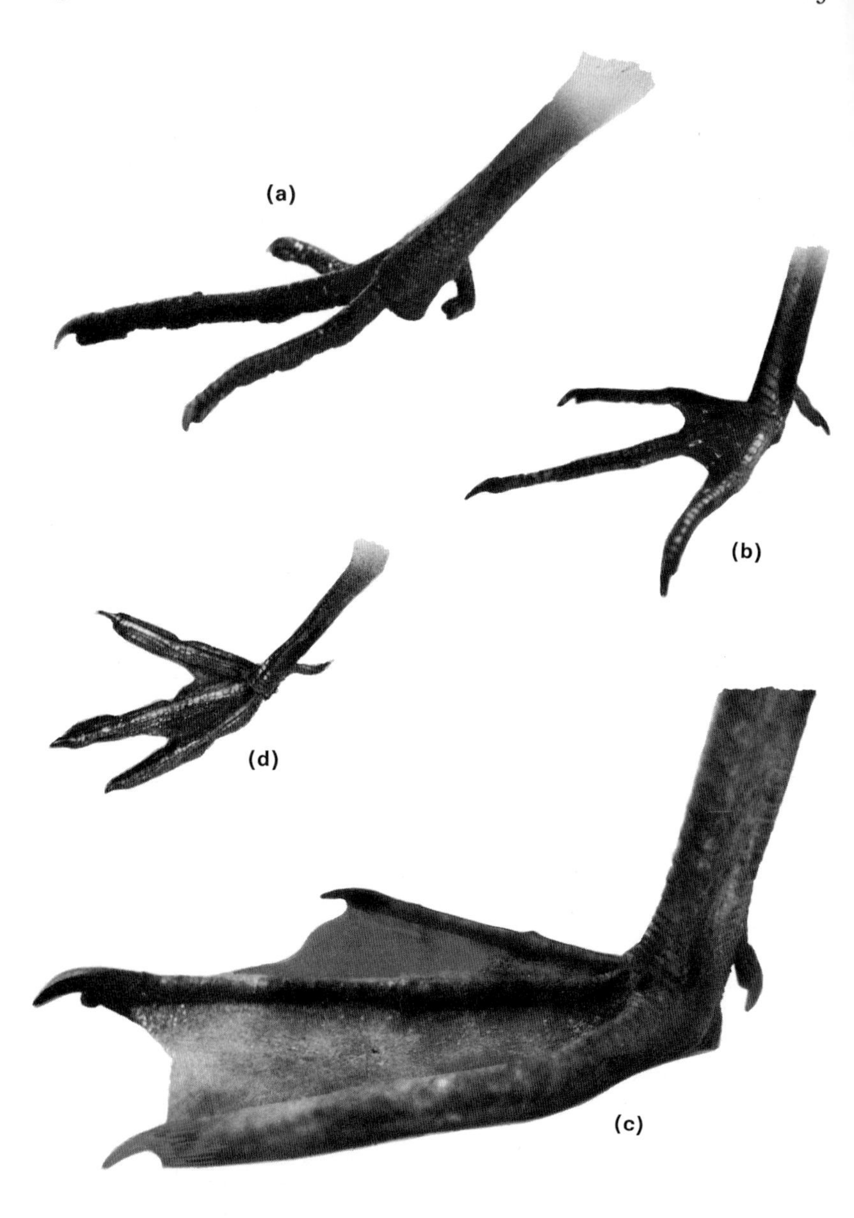

Fig. 3.5 The feet of (**a**) a woodcock (**b**) a redshank (**c**) a glaucous gull (**d**) a red-necked phalarope.

of their insect-eating relatives to a different type of prey. Other birds have gone a little further, and habitually swim on the surface but do not go beneath. In the order Charadriiformes the avocets (*Recurvirostra*) and phalaropes (Phalaropidae) are slightly specialized waders that do so, and the gulls and skuas (Laridae) spend much of their time in that way. A similar and perhaps greater range is found in the Gruiformes, in which are generally included highly terrestrial and cursorial forms such as the cranes (Gruidae) and bustards (Otididae), as well as the rails (Rallidae), some members of which, such as the moorhen (*Gallinula*), spend as much time on water as on land.

The only obvious adaptation to their partially aquatic mode of life shown by most of these birds is a greater or lesser degree of webbing of the feet. There are two main variants: a web that includes the second, third and fourth toes only; and separate webs for each toe (Fig. 3.5). These conditions have obviously been achieved independently and more than once, and illustrate the point made above, that superficial characters are easily acted on by natural selection, and are of no great importance in classification.

The two orders just quoted have also members that habitually dive and obtain their food under water; the coot (*Fulica atra*) in the Gruiformes, and the auks (Alcae) in the Charadriiformes. There are also some orders all of whose members dive (Impennes or penguins; Gaviiformes or loons—the term 'diver' also applied to this order is ambiguous in the present context—and Podicipediformes or grebes) and some in which all the members are aquatic, all of them swimming but some rarely or never diving (Anseriformes or ducks and geese; Pelicaniformes, including pelicans, cormorants, gannets and some others; and Procellariiformes or petrels).

The diving birds amongst these eight orders can be divided functionally into those that, when under water, swim with their feet, and the penguins, petrels and auks, which use their wings; there are a few partial cases of the use of both, for example some shearwaters and loons, but the main method is always clear.

The foot-divers have webs, which may be of either of the types found in the surface-swimmers, and some have a third type, in which all four toes are included in the web (Fig. 3.6). The whole leg is often much more modified than in the swimmers.[348] The legs are generally far back, which gives an upright stance and awkward gait on land (well seen in the grebes), but also allows much greater freedom in the direction of the swimming stroke. Combined with an ability to rotate the tibiotarsus, it enables both grebes and loons to steer by their propulsive strokes. The muscles of the leg and their insertions on the tibia are large. Independent evolution is shown by the different ways in which the increase is achieved; different muscles are used

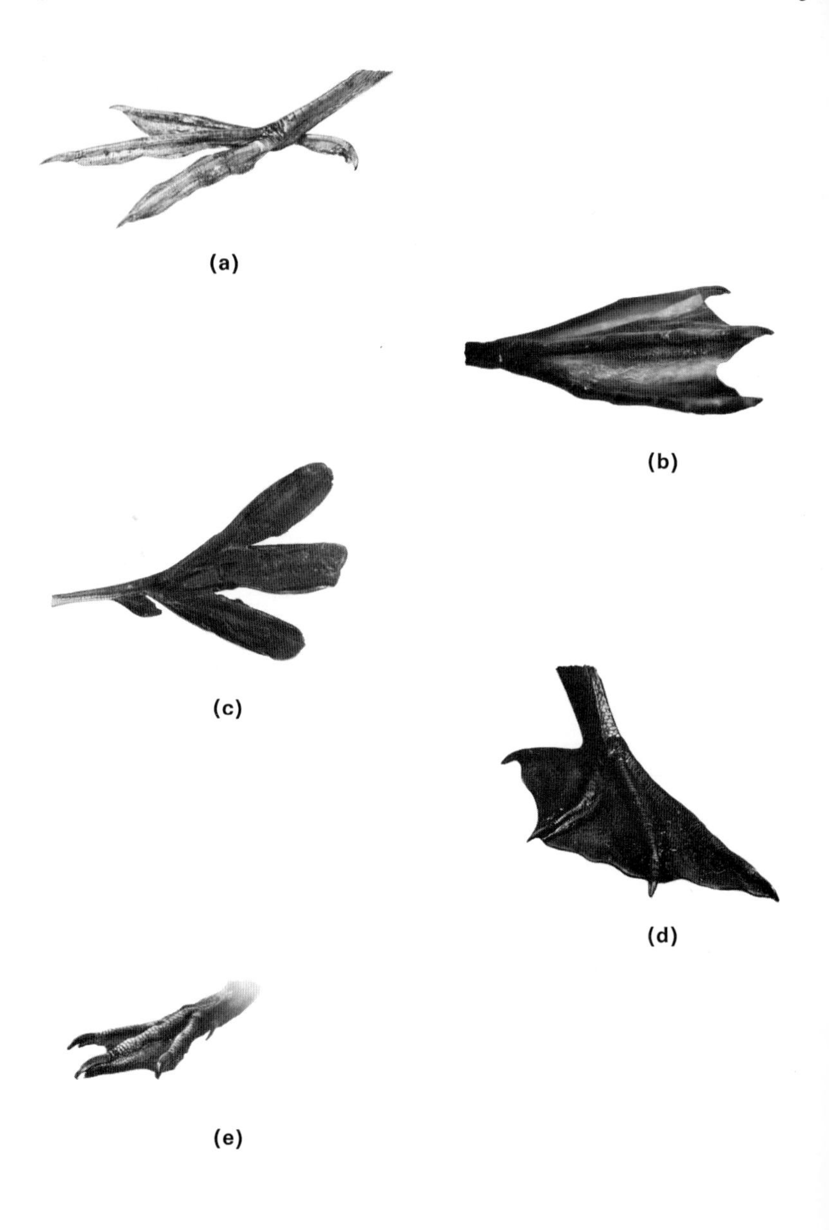

Fig. 3.6 The webbed feet of (**a**) a coot (**b**) a guillemot (**c**) a great crested grebe, the right foot seen from below (**d**) a cormorant (**e**) a penguin.

in grebes and loons, and only in the former does the patella contribute to the very large cnemial crest (Fig. 3.7).

Wing-divers are all marine, and for the most part feed well away from land. The petrels and most of the auks, in spite of some modifications in

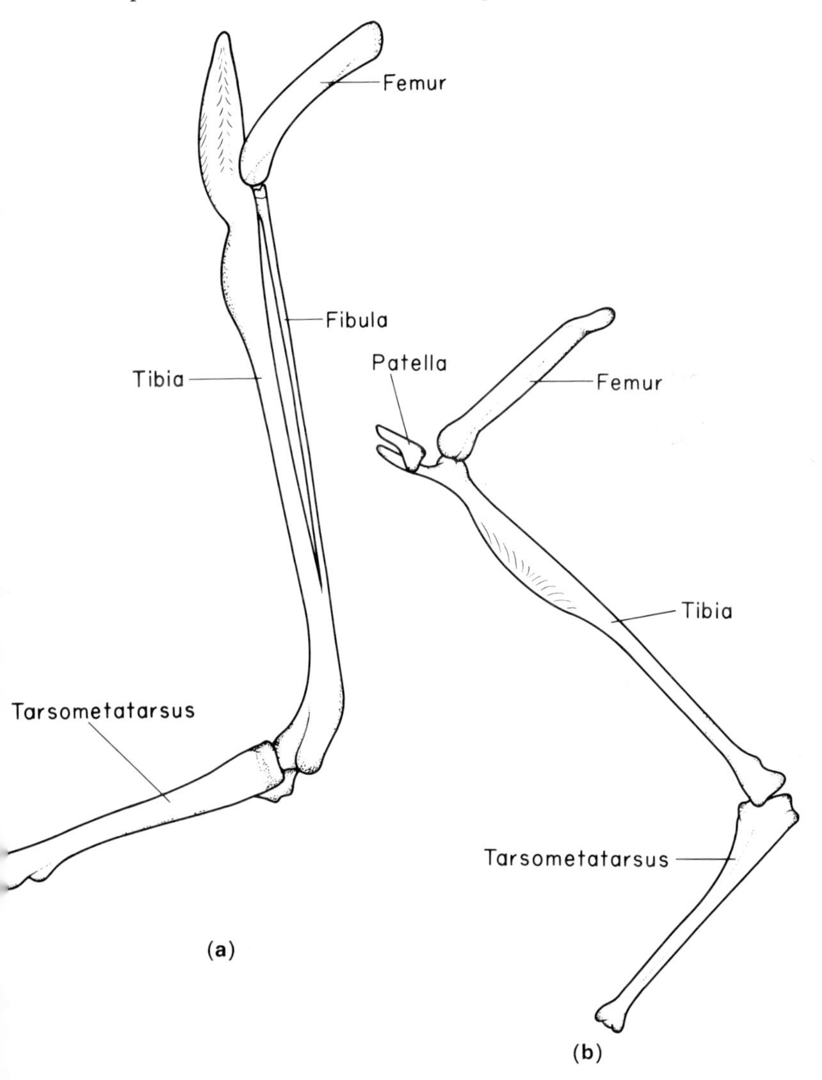

Fig. 3.7 Bones of the left leg of (**a**) *Colymbus immer*, ×1.0 and (**b**) *Podiceps cristatus*, × 3/5.

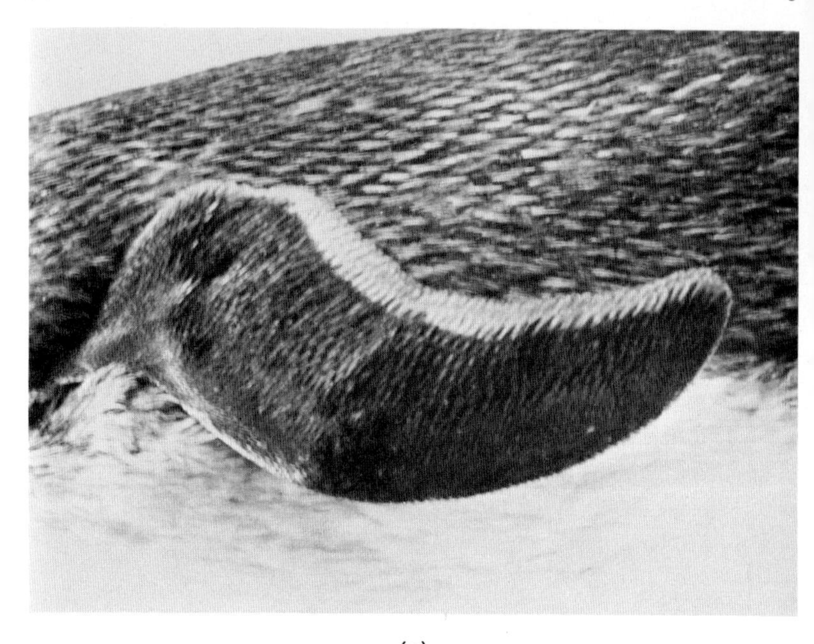

(a)

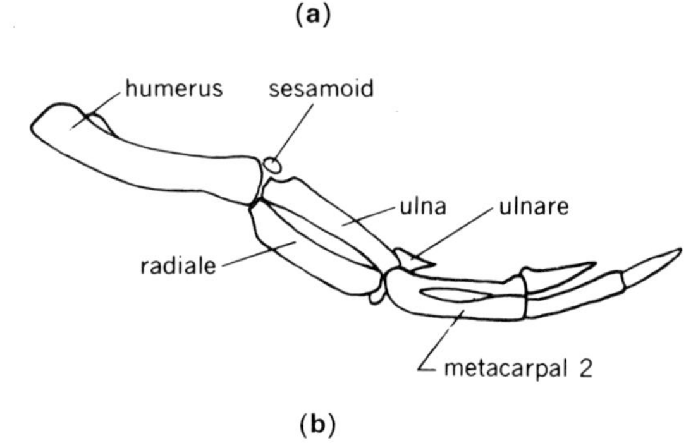

(b)

Fig. 3.8 (**a**) Wing of a penguin. (**b**) Skeleton of the wing of a penguin, ×1/3. ((**b**) from W. B. Yapp, 1965, *Vertebrates: their structure and life*, Fig. 17.23, Oxford University Press, New York.)

the wings, can fly well; the breeding puffin (*Fratercula arctica*) commutes daily from nest to feeding waters for distances of as much as 50 miles. Its peculiar flight, with short strokes, is presumably imposed on it by the requirements of using its wings under water, where a long stroke would be difficult. The medium-sized auks, such as the razorbill (*Alca torda*), do not need the full surface of their wings in diving, for during the moult, in which all the flight feathers are lost together, they can swim submerged but cannot fly. The great auk had only small wings and could not fly. As would be expected, the legs of wing-divers are little modified.

The penguins also are flightless, and have a highly modified wing which they use for swimming even on the surface (Fig. 3.8). The distal musculature, which in flying birds is highly important, has almost disappeared, so that the wing is little more than a flipper like that of a seal or plesiosaur. It is stiffened by being filled up with greatly flattened bones; there is no trace of the hyperdactyly or hyperphalangy found in comparable limbs in reptiles and mammals. There are no remiges.

The dippers (*Cinclus*), of which there are four closely-related species in the northern hemisphere, are passerine birds that have taken to an aquatic life with no very apparent modifications in structure. They swim on the surface using their feet, which are not webbed, and under the water using their wings. These are short, like those of the wrens, to which the dippers are possibly related, but the bird spends much time on the wing, and its flight does not seem to be handicapped.

It is often said that diving birds have of necessity increased their specific gravity by having less air amongst the feathers and by reducing the pneumaticity of the bones, but the evidence is not conclusive. If the water in which a duck is floating has a good detergent added to it, so that the feathers are wetted, the bird sinks at once, showing that its overall specific gravity of less than one is produced by the air in its plumage.

3.23 Typical birds

There remain the birds which live what may be called a typically bird-like life, spending much time on the wing and living for the most part amongst trees or bushes. Even in these there are specializations. The Ciconiiformes (herons and storks) feed mostly in water but seldom or never swim, and are chiefly notable for their long legs. The Falconiformes (hawks, falcons and vultures) are diurnal birds of prey, strong of flight and with adaptations in strong claws and hooked beaks for attacking other animals (Fig. 3.9). There is a growing feeling that the group is polyphyletic. The Strigiformes (owls), with similar adaptations of beak and claw, have long been recognized as having no close connection with the hawks. They are mostly nocturnal, and one of their most interesting adaptations is a great development and asymmetry of the ear, which enables them to determine the exact position of the small mammals on which they feed by the squeaks that they give out (Fig. 3.10). The Caprimulgiformes (nightjars) have some similarities to the owls in their feathers and their egg-white proteins, but are very different in other respects, such as the feet. The typical Piciformes (woodpeckers) climb trees and have strong claws and

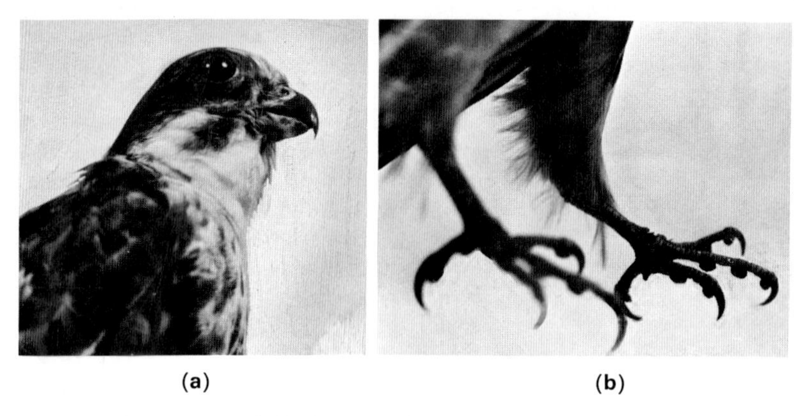

Fig. 3.9 (**a**) Head, and (**b**) feet, of a hobby *(Falco subbuteo)*.

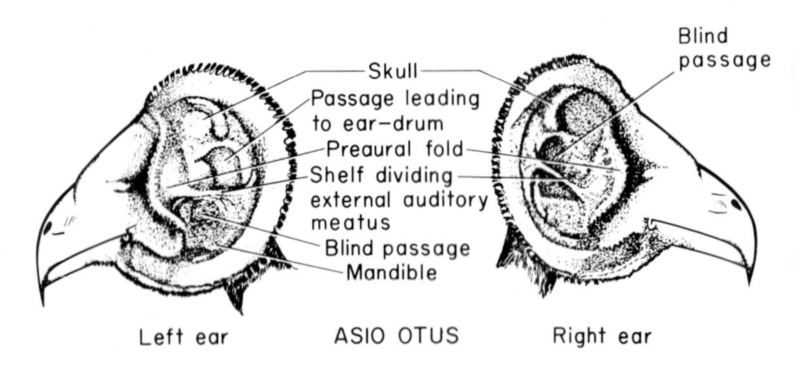

Fig. 3.10 Diagram of the ears of a long-eared owl, *Asio otus,* to show the asymmetry; the preaural fold has been turned forward to show the opening of the external auditory meatus. In the left ear the blind passage is below the shelf, in the right ear it is above. (From Pycraft, 1898, Plate 27, *Trans. Linn. Soc. Zool.* (2) **7**, 223.)

stiff tail-feathers. Tree-climbing has also developed in some birds of other orders.

The Apodiformes (swifts and humming-birds) have concentrated on flight, and have short legs with weak toes, so that they are unable to do more than shuffle when on the ground. The humming-birds are mostly very small (*Mellisuga helenae* of Cuba has a body only an inch long) and vibrate their wings very quickly, so that they have some convergent similarities to large insects. They are the only birds that can fly backwards. They are a successful group, with more than 300 species.

The Psittaciformes (parrots) are a uniform and peculiar group which superficially look a little like birds of prey. But the long claws and hooked bill are used not for predation but for clambering about in trees, the beak being used as a third hand; the bird can even hang by it momentarily. Food is held in one foot while it is nibbled by the beak and thick tongue. Both these features are unique amongst birds. The brain is large and there is some degree of binocular vision.

Finally there are the Passeriformes, the perching birds or song-birds (although strictly the last name applies only to the sub-order Oscines), which include more than half the total number of species.[31] Their classificatory feature is the great development of the organ of voice, or syrinx (Fig. 3.11). This is a specialization at the point where the trachea divides

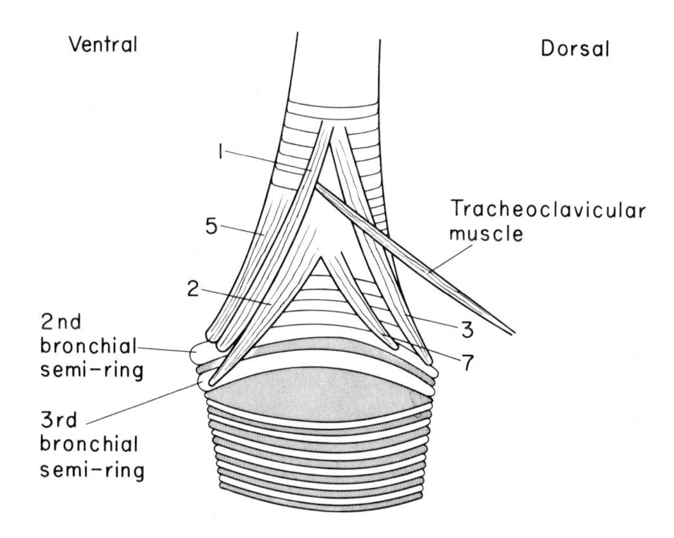

Fig. 3.11 Syrinx of a raven, lateral view. In the syringeal muscles (Nos 1–7), 4 is covered by 3, and 6 by 2. (Redrawn from Gadow, in Newton, 1896, *A Dictionary of Birds*, p. 941, A. and C. Black, London.)

into the two bronchi, generally formed from all three of these tubes, and differs both in structure and position from the larynx of mammals. The acoustical principle however is the same; membranes are made to vibrate by the passage of air over them, and the tension in them, and so the note that they give out, can be varied by the contraction of muscles. The great feature of the Oscines is the number of these muscles; they have from four to nine pairs, five or seven being the commonest. Only parrots and lyre-birds (Menuridae, Passeriformes) have three pairs, a few species have two, most birds only one, and the ratites and a few others, none.

One would expect the complication of sound that a bird can produce to have some relation to the complexity of its syringeal muscles, and in a general way this is true. Many Oscines have songs with a range of pitch and quality found in no other animal sounds except the voice of a trained concert or opera singer, and in the ability to produce complex sounds, grace-notes and trills, they probably surpass even these. But the possession of many muscles does not necessarily mean an ability to sing; crows, for example, have seven pairs, but have a very limited voice. On the other hand parrots, which like the crows make few and harsh sounds in nature, are able to mimic the human voice, even sometimes to its inflexions and local accents.

It is customary to regard the Oscines as the 'highest' birds, but the term is seldom defined. Newton said 'those which are best furnished with a brain are superior to those which are less well endowed', and concluded that on this basis the Passeriformes was the highest order and that the crows were the highest passerines.[5] It certainly seems that some passerines, such as finches, crows and tits, can learn new habits, such as opening milk bottles and pulling up food with a string, in a way impossible to others, and crows appear to be better than pigeons at counting. On relative size of brain, however, the leading birds so far as is known are the parrots. It is noteworthy that their habit of holding food and other things in front of them is the same as is found in the primates, where it is generally considered to have led to stereoscopic vision and the large brain of man. Perhaps lists of the orders of birds ought to end with the Psittaciformes.

3.3 SPECIES AND SUBSPECIES

No satisfactory definition of the term 'species' has been produced since men abandoned literal belief in Noah's ark. All those current in the text-books emphasize an interbreeding population and so break down both logically because the males or females cannot breed with each other, and semantically because they exclude asexual reproducers such as *Amoeba* and non-sexual forms such as neuter insects. But every zoologist and every naturalist has a working idea of what he means by 'species', and uses the idea in his writing and reading. All would probably agree that two forms of an animal that live in the same geographical area and regularly interbreed belong to the same species, however much they differ in appearance. Differences of opinion arise where two forms that differ little are separated geographically, so that they have no chance of interbreeding, and we cannot tell what would happen if they were to meet.

Ornithologists have been prominent in developing the concept of a smaller unit, the subspecies, a subdivision that may be regarded as an incipient species. It has been defined by Mayr as 'a geographically localized division of the species, which differs genetically and taxonomically from other subdivisions of the species'.[243]

So defined, it is unobjectionable, but one may ask whether there is any advantage in describing a chaffinch from Europe by a different trinomial name (*Fringilla coelebs coelebs*) from one from Britain (*Fringilla coelebs gengleri*), when the country of origin will be stated anyway. One of the difficulties is that though we can be sure that almost any two populations we like to choose will have some genetic differences, we seldom have much information on what they are, or on how far apparent differences are genetic and not environmental, or on how far the differences are enough to cause reproductive isolation. The biggest differences between the British and continental chaffinches are not in colour but in voice, both call-notes and song; in particular, Scandinavian birds add a distinct note, similar to the call of the great spotted woodpecker, at the end of the song. Since the song of the chaffinch is learnt, it has been fairly generally assumed that these dialects cannot lead to isolation and so have nothing to do with incipient speciation. The conclusion is rash; mated hens can recognize the song of their own cocks and ignore that of others, so the possibility of a restricted selection of mate is there.

On the other hand, the slight differences in colour, and especially the slight differences in dimensions (very seldom properly treated statistically), on which most subspecies are based, are not necessarily genetic in origin and have seldom been proved to be so. In the 1920s and 1930s there were obvious marked differences in height between the social classes in England; these may have been partly genetic, since there are well-marked differences in height between races of men that are undoubtedly genetically distinct, but experiments with dietary supplements (especially milk) showed that most of the differences could be removed in this way. Further, we know that natural populations, for example the Icelanders, have changed their stature, and reversed the change, as their diet has altered.

Even characters that are undoubtedly genetic can change rapidly under selection. We know this best for Lepidoptera; for example populations of the meadow brown butterfly (*Maniola jurtina*), which differ in colour to a degree that most ornithologists would regard as justifying subspecific distinction, have been shown to change quite rapidly.[71] We know that the colour is inherited, and that selection acts through an inherited response to drought or parasitism which is linked with it.

The problem of subspecies is made more difficult by the existence of clines, which are gradients of a character following geographically across the distribution of a species. They are of two sorts. First, there may be a

(a)

(b)

Fig. 3.12 Heads of (**a**) a normal, and (**b**) a bridled guillemot.

gradient, for example of colour, so that, though the exact provenance of a specimen cannot be determined, the approximate region from which it came is usually pretty clear. The nuthatch (*Sitta europaea*), which is divided into many subspecies, is in general pinkish buff on the breast in western Europe, less so as one goes east, white in eastern Russia, and then

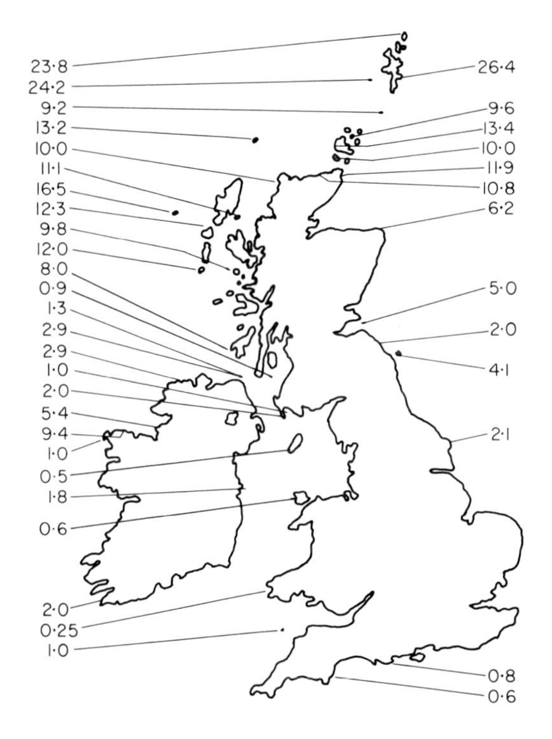

Fig. 3.13 Map of the British Isles showing the percentages of the bridled form of guillemot at breeding colonies in 1936–41. (From Southern and Reeve,[335] 1941, Map 1, *Proc. zool. Soc. Lond.* **111**A, 264.)

buff again in China. Secondly, there may be two clear-cut varieties, which are distributed in different proportions in different places. The common guillemot (*Uria alge*) has a form called 'bridled' in which a white line runs round the eye and extends backwards (Fig. 3.12). The frequency of this alters from north to south[335] (Fig. 3.13).

As Tucker pointed out twenty years ago,[370] there is no justification for using subspecific names unless the differences are clearly visible. In the field they seldom are, so at best trinomials should be confined to the museum. Nor are they of much use when we know nothing of the genetics or stability of the characters involved. Much that has been written about subspecies in birds seems to have been produced in ignorance of elementary genetics or of what is known about other animals. Some island subspecies are probably no more than the chance result of colonization by a small number of individuals; this is the most economical explanation of the origin of the Scottish crossbills (*Loxia curvirostra*), whose breeding colony is probably only a century and a half old.

The assignment of a bird to a subspecies, and so to a particular part of the world, can lead to wrong conclusions. The Irish coal tit (*Parus ater*) is certainly yellower than most of those in England, and is usually separated as *Parus ater hibernicus*. But some degree of yellowing is present in many English skins, and some English birds are as yellow as any from Ireland.[407] Whatever the cause of the variation, the trinomial is not helpful, and to say that a yellow bird was Irish (as most ringers would do) is about as wise as saying a man is Irish because he has red hair. Use of Latin trinomials is perhaps most justified when there is a clear-cut colour difference which is known to be inherited and which is broadly defined geographically. The best example is the carrion crow (*Corvus corone corone*) with its variant the hooded crow (*Corvus corone cornix*). There is a fairly sharp line of demarcation between the breeding ranges of these two (Fig. 3.14), and where they meet they regularly interbreed.[245,344] The genetics of the crosses are not fully known, but it is clear from the nature of the mixed broods, which can range from one type to the other with all possible intermediates, that several genes are involved and that the types are not always homozygous.

Subspecies are incipient species, and it seems likely that they cannot develop into the latter without some degree of isolation.[243,255] When this is clear-cut and geographical, as with resident populations on islands, speciation can probably be rapid. The Canary Isles, which geologists say were formed in the Pliocene some million years ago and have never been connected to the mainland of Africa, have three endemic species out of a total of 46 residents.[26] Many islands have subspecies which must have been formed since the last glaciation, about 20,000 years ago. Birds are, however, very mobile, and isolation is likely to be occasionally upset, especially in migratory species. There is a strong tendency for individuals to return to their birthplace, but in general, not only in birds but throughout the animal kingdom, the more mobile groups have fewer subspecies. Ducks, in which pairs are formed in the winter quarters where populations are mixed, have fewer subspecies than geese, where the pairs are not formed until the birds have returned to their breeding place. Populations of the song-sparrow (*Melospiza melodia*) living in salt-marshes in California, which are highly sedentary, have possibly formed subspecies in 50 years.[175]

Most ornithologists believe that the only sort of isolation that can lead to speciation is geographical, but the case is not proved. On the Canary Isles of Gran Canaria and Teneriffe there are two sorts of chaffinch, *Fringilla coelebs tintillon* (the same species as in Europe and the mainland of North Africa but a different subspecies) and the blue chaffinch, *Fringilla teydea*. The orthodox explanation is that there have been two invasions of the islands by birds from the mainland. The first formed a population which was isolated long enough ago for it to have formed a separate species, so

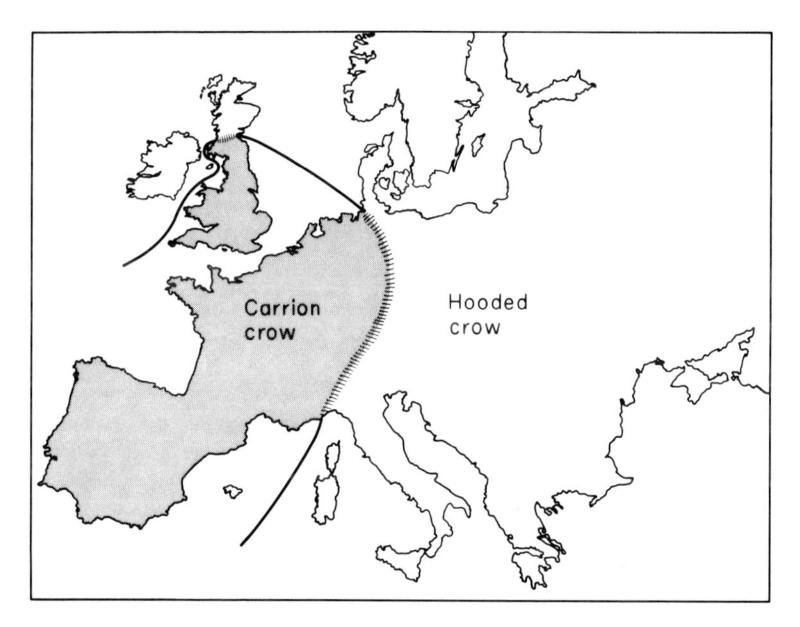

Fig. 3.14 Map of Europe showing the distribution of carrion and hooded crows. The hatched line shows the zone of interbreeding.

that when the second invasion took place the two sets of birds were reproductively isolated and no mixing took place.[198] But there are other facts to be considered. Both *F. teydea* and *F. coelebs tintillon* are much bluer than other chaffinches, including those from nearby Africa, and their songs and calls also resemble each other. *F. teydea* is restricted to the tops of the mountains and the pine forests found there; *F. coelebs tintillon* is, unlike most chaffinches, not found in the pines, or only rarely. An alternative explanation therefore is that there was a single invasion, and that those individuals that became resident in the pine forests evolved into what is now regarded as a distinct species.

The main point to be made is that we do not know. We do not even know whether *F. teydea* is really reproductively isolated from other chaffinches, or indeed whether *F. coelebs tintillon*, which was formerly regarded as a good species under the name of *F. tintillon*, is similarly isolated from mainland birds. Speculations about the origin of *Fringilla teydea* must not be used, as they often are, to support the hypothesis of 'no speciation without geographical isolation' on which they are based, for such an argument is circular.

4

Physiology

We know much less of comparative physiology than we do of comparative anatomy, but enough work has been done on scattered members of the main vertebrate classes to show that the functioning, as well as the structure, of their organs and tissues is fundamentally the same. Nevertheless there are differences from one animal to another, and we might expect that birds, which live a type of life very different from that of most other vertebrates, would have some peculiarities of their own. They certainly do, but unfortunately the only birds that have been even moderately well studied by experiment are domestic poultry, which are somewhat specialized both in habit and in structure, and generalization is both difficult and dangerous.

4.1 NUTRITION

4.11 Food

The food of birds, like that of other animals, consists chiefly of carbohydrates, fats and proteins, and the list of essential aminoacids is the same for the fowl as for man. Vitamins have been little studied, but it is probable that all or most of those used by mammals are used by birds, and must be supplied or synthesized. The early work on the necessity for vitamin B was carried out largely on pigeons. Fowls can synthesize nicotinic acid from tryptophan, and some birds, probably most, can synthesize ascorbic acid. Fowls have a high requirement for vitamin D, and can use it only in the form of cholecalciferol. House-sparrows and starlings need much less, possibly because they obtain it by irradiation of the oil in the preen gland, which contains cholesterol. Even less is known of the need for trace-elements, but manganese appears to be especially important both in fowls and in ducks, for healthy growth and for egg-production. Species of mammals differ in their quantitative requirements of trace-elements, and soils and plants differ in the degree to which they can supply them, so that it is possible that some of the peculiar distributions of animals might be determined in this way.

4.12 The alimentary canal

The alimentary canal of birds has the usual vertebrate parts: buccal cavity, oesophagus or gullet, stomach, small intestine divided into an

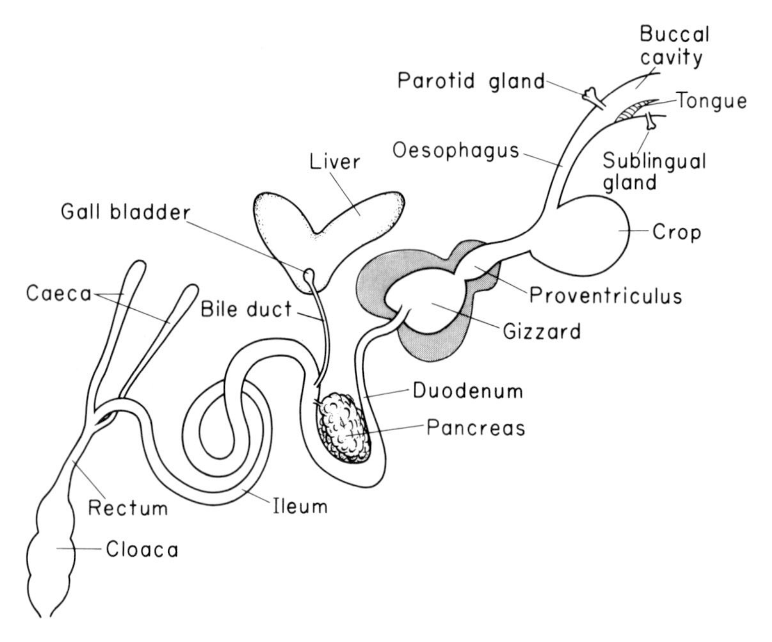

Fig. 4.1 Diagram of an alimentary canal of a bird.

anterior and glandular duodenum and a posterior and absorptive ileum, and large intestine or rectum (Fig. 4.1). Salivary glands open into the buccal cavity, and pancreas and liver into the duodenum. At the junction of the ileum and rectum are two caeca, or occasionally, as in grebes and herons, there is only one.

There are two peculiar features which, because they are present in pigeons and game-birds, the only birds of which most people ever see the inside, are generally thought to be characteristic of birds, but which in fact are found only in a minority. The first of these features is the crop, a thin-walled expansion of the lower end of the gullet, which is well-developed only in seed-eating birds such as finches and buntings, game-birds, parrots, and pigeons. A narrower crop, sometimes only temporary, is present in some fish-eaters and carnivores, and in the humming-birds. In all these its function seems to be simply that of storing food which is temporarily abundant but cannot for the moment be digested. A wood-pigeon (*Columba palumbus*), for example, may swallow several acorns in quick succession, and gradually release them for digestion. As many as 61 have been recorded in the crop at one time.

In two groups of bird the crop has acquired an additional function. In the hoatzin its walls are thick and muscular, and squeeze the juice out of the leaves on which the bird feeds. In pigeons during the breeding season the cells of the walls of the crop in both sexes break down, and so liberate into the lumen material, of the consistency of soft cheese, which is rich in casein and contains also fat and lactose; its composition is thus very similar to that of milk. It is regurgitated and fed to the young, and is appropriately called pigeon's milk. It has, however, no relation to the milk of mammals beyond its chemical composition and the fact that its production is stimulated by prolactin, a hormone secreted by the pituitary gland.

The second feature is that most seed-eaters, and some others, have the stomach divided into two, an anterior glandular proventriculus, and a posterior muscular and grinding gizzard. In some birds, chiefly fish-eaters such as herons, but also most of the ratites, there is a third division of the stomach, the pyloric bulb, constricted from the posterior part of the gizzard.

The relation of the length of the intestine to the food is not as clear as it is in mammals, possibly because of the complications introduced by the relative length of the caeca. In general it is relatively short in frugivorous and insectivorous species, and long in those that feed on fish, seeds or leaves.[345] The intestine of a carrier pigeon is one metre or more long; that of the long-eared owl (*Asio otus*), a bird of about the same size, is only half this length. The bullfinch, which has an almost completely vegetarian diet, has an intestine twice the length of that of the chaffinch, which is the same size but eats some insect food. In both the caeca are vestigial. The fowl has long caeca, the pigeon, with a very similar diet, has short ones. Within the order Galliformes the browsing species have longer intestines and caeca than those that eat seeds.[207]

4.13 Digestion[345]

The salivary glands of the fowl produce a saliva, which, like that of man, is probably primarily lubricating. It contains an amylase, but birds bolt their food, and there will be no time for an enzyme to act before it is swallowed. In the crop, however, it may have some effect, and grain in the crop of pigeons is often considerably softened, either by this salivary amylase or, more probably, by bacterial and autolytic action. In the fowl such action has been shown to produce lactic acid. The crop walls produce no enzymes.

Saliva also contains mucin, and a number of birds use this in the construction of their nests.

Many birds eject from their mouths the indigestible parts of their food, whether bones, fur, chitin or thick cellulose.[369] Pellets of such material are produced by all carnivores, most piscivores, probably all insectivores, and some vegetarians such as rooks (*Corvus frugilegus*). It seems to be generally assumed that the pellets come from the crop, but at least in the carnivores it is more likely that they come from the proventriculus. Owls generally swallow their prey whole, and the pellet is a tightly-packed mass of fur and bones, with little flesh. It is difficult to see how this could be produced except with the help of a proteolytic enzyme, and none is produced anterior to the stomach. The crop is in any case not well developed, but secretions from the proventriculus may travel forward into it. The ejection of the pellet takes place by a reversal of the ordinary muscular movements; in the great horned owl (*Bubo virginianus*) it is preceded by nausea for 15 or 20 minutes, during which it can be induced by gentle pressure on the abdomen, but in other species it seems to occur suddenly.[298]

In insectivorous species ejection is seldom observed, and the pellets are too small to be found unless they are collected immediately, but the birds of prey (raptors) usually have favourite stations for throwing up their large pellets, so that they are easily found. If they are carefully pulled apart the bones they contain can be identified, and the species and the minimum number of prey present determined. Often the remains of only one or two individuals are found in each pellet, but there is some relationship to the size of prey. Those of a fairly long series of the barn owl (*Tyto alba*) contained an average of 5·2 skeletons in Wiltshire, where the prey was mostly voles and shrews, but only 2·2 in Norfolk, where many young rats were taken.[366] The largest number of skeletons found was nine. In short-eared owls (*Asio flammeus*) in Scotland, feeding exclusively on voles and shrews, the average was just over three, but counts in North America suggest that in that continent one is the rule in this species.[97] Three and two were the commonest numbers for a series from the long-eared owl in England,[366a] while in a shorter series from Iraq one was usual.[153a]

Owls in captivity take only one or two meals in the course of 24 hours, but in the wild they must take more; probably in general one or two pellets are produced in this time.

Many species regurgitate food that has been swallowed, and sometimes partly digested, to feed to their mate or nestlings. Gulls, for example, which have no crop, feed their chicks with material that is sometimes so much attacked as to be unrecognizable; it must presumably come from the proventriculus. Parrots, which have a crop, similarly use seeds, which have been only slightly softened.

The walls of the proventriculus produce a peptic enzyme, but the strength of this, and the acidity of the lumen, vary widely, as they do in mammals. The pH ranges from 2·5 to 4·5, and, as would be expected, pepsin is strongest in carnivores; in the great horned owl its activity is greater than that in the dog, while in fowls and pigeons it is low. As is usual in vertebrates other than mammals, acid and enzyme are produced by the same cells.

After peptic digestion the food passes to the gizzard, if there is one, and is subject to considerable trituration, sometimes assisted by swallowed stones. Both in the fowl and the pigeon the gizzard is more acid than the crop, which implies that the walls secrete hydrogen ions, but there is little digestion here. A fowl from which the gizzard has been surgically removed can digest finely-divided corn as well as can an intact bird, but it does not do so well with whole seeds, which will be its natural diet. It is for these that the grit in the gizzard is most useful. In the domestic goose there is much carrying forward of secretions from the intestine, and the food is apparently shunted backwards and forwards between the two parts of the stomach.

The walls of the gizzard produce a secretion which sets to form a protective cuticle; it is continually worn away and in some birds, such as the starling (*Sturnus vulgaris*) and mistle-thrush (*Turdus viscivorus*), it is periodically cast off and vomited through the mouth. Hornbills (*Buceros* and other genera) similarly eject the endothelial lining of the gizzard as sacs containing seeds that the male probably feeds to the female on the nest.[265]

The secretion of enzymes and the digestion of food in the small intestine are generally similar to those in mammals. The pH rises to near neutrality, and all three classes of foodstuff are digested. Where caeca are present, the food passes into them and digestion continues. Chitinase has been described from the proventriculus of a few insectivorous passerine birds, including the house-sparrow and blackbird (*Turdus merula*), but could not be found in pigeons.[174] The honey-guides (Indicatoridae) which feed largely on bees-wax do not produce their own esterases, but depend on a micrococcus and a yeast in the gut. Both can split esters of the wax but the two together do better than either species alone.[116] So far as is known, symbiotic microorganisms are not important in birds in the way in which they are in many herbivorous mammals.

The control of secretion seems not to be so fine as in mammals. In both fowls and pigeons gastric secretion is intermittent and under the control of the nervous system. In fowls secretion of acid can be provoked by 'sham feeding', for example with water; the stimulus seems to be the mere distention of the gullet, and not, as in mammals, the presence of material of a particular chemical nature. Secretin is present in the intestine of the fowl (and in that of many other vertebrates) but its function in any except mammals is obscure. In the goose there is no regular relationship between the secretion of bile and pancreatic juice and the exit of the food from the stomach. In most small birds food passes through the gut in a few hours.

Clearly birds absorb the products of digestion, but little work has been done to find out how they do it. Bile salts, in the dozen species that have been examined, are similar to those of mammals, and presumably help in the digestion and absorption of fat.[155] Although fat makes a small part of their diet, fowls absorb about 90 per cent of that supplied. The order of ease of absorption of the hexoses is the same as in mammals; galactose is the best, and then glucose, xylose and fructose. For the first two of these active transport has been demonstrated.

4.2 METABOLISM

When the food has been absorbed into the blood it has to undergo further chemical changes, and we know very little of these in birds. As before, we can say that while the general lines of change are similar to those in mammals, there are important differences. Some of these are probably adaptive, while others may derive from an early stage in the vertebrate ancestry.

4.21 Biochemistry of carbon

The glucose in the blood of the few species of bird that have been examined has about twice the concentration of that in mammals, which may be connected with the higher metabolic rate of birds.[92] It is controlled in some way by the secretions of the islets of the pancreas, but there are some anomalies. In general, the pancreas is rich in *A*-cells (which in mammals produce glucagon, which raises blood sugar) and poor in *B*-cells (which produce insulin, which lowers blood sugar), so that hyperglycaemia would be expected.[157] In the fowl injection of glucagon raises the amount of glucose in the blood, and greatly increases its content of free fatty acids. Insulin causes a decrease in glucose, and a large increase of free fatty acids, which may be due to the fact that insulin also causes the release of glucagon. In mammals, both hormones lower the free fatty acids in the plasma. Removal of the pancreas from an owl caused extreme hyperglycaemia and death, just as it would in a mammal, but in grain-eating domestic poultry it causes only a mild increase in blood sugar. The blood sugar of a pancreatectomized duck increased still further when the animal was fed with meat.

Reptiles also are relatively insensitive to insulin, and the identity of their *A*- and *B*-cells with those of mammals is not certain. It looks as if the separation and importance of insulin and glucagon were not fully established at the time when the cotylosaurs branched into the synapsid and archosaurian lines, and that birds are slowly and independently evolving something similar to what has been achieved by mammals, but with a greater dependence on fat as the conveyor of energy.

4.22 Biochemistry of nitrogen

The history of the nitrogen metabolism of birds is different from that of carbon metabolism. They do not have the ornithine cycle, and produce only a little urea, probably by the action of arginase on arginine derived by hydrolysis of proteins in the food. Since the ornithine cycle is present in all the other main groups of living vertebrates except the teleosts, birds have almost certainly lost it. The ammonia separated off from the amino-acids undergoes a series of reactions with carbon dioxide, glycine, formate, glutamate and ribose-1-phosphate to give hypoxanthine (6-hydroxypurine). Small quantities of this substance are formed from nucleoproteins by mammals, and it is then oxidized to uric acid. It follows the same route in birds; they have no uricase, and do not oxidize it further, while most mammals (but not man) convert it to allantoin.

Guanine, which is such an important constituent of guano, the accumulated excretory product of tropical sea-birds, is a stage in mammals in the conversion of nucleic acids to uric acid.[179] In the birds it is derived from their fishy food; if they are fed on meat they do not produce it.

The advantage of converting the excretory nitrogen to uric acid is that it can first be formed as a soluble sodium urate, and then precipitated as the free acid, so that most of the water can be resorbed. The urine of birds is generally described as being 'semisolid'. It has much the consistency of a soft toothpaste, and is formed by the gelation of a supersaturated solution of uric acid. The respiratory water loss of birds varies inversely as some function of body weight, so that birds weighing about 80 g or more have no difficulty in replacing the loss entirely by metabolic water. Any saving in the urinary loss will assist them. Many insects produce a similar urine, also containing uric acid, and for the same reason.

4.23 The kidney[60]

The working of the kidney does not seem to differ much from that of mammals. Water and crystalloids are filtered from the glomerulus under pressure of the heart-beat, and water and some substances are absorbed back into the blood from the tubule. Birds produce very little urine, and almost all the water filtered by the glomerulus must be absorbed. Approximate calculations show that in the fowl it must be about 99 per cent,[319] but as the more precise information available for man shows that he absorbs more than this, the figure is probably of little value. The amount of water absorbed depends on the intake, and is, as usual, under the control of the antidiuretic hormone of the hypophysis. Other substances are absorbed as well, and the tubule is secretory, adding waste products to the urine. More than 90 per cent of the uric acid excreted is lost in this way.

The kidneys of some birds seem to be incapable of dealing with high salt loads, and in many sea-birds, including penguins and cormorants, the sodium chloride that they absorb is eliminated through the nasal glands.[311, 312] Experimentally this secretion can be increased by the injection of cortisol, one of the products of the adrenal cortex. Many species of hawks also lose much sodium chloride from the nostrils, in a fluid hypertonic to the blood, though their potassium is lost chiefly in the urine.[53] Presumably their diet of flesh and blood gives them a high intake of salt. In contrast, many finches can produce a urine hypertonic to the blood, and can take considerable quantities of salt in the food. The crossbill eats it if it is available, and the savannah sparrow (*Passerculus sandvichensis*), which lives in salt-marshes (among other places), can raise its urinary chloride to seven times that of the plasma.[79]

4.24 Production of energy

The large amounts of energy needed in flight have been referred to in Chapter 2, but even apart from this birds in general have high metabolic rates, which may be connected with their high activity, and their high temperatures (discussed more fully below). Oxygen consumption per unit mass rises as the mass become less, owing to the greater proportionate heat loss in small animals. It is probably a mistake to try to fit all birds to one equation, since there must be variables that cannot be allowed for, such as the thickness of the feathers, but in general it seems that, as in mammals, the resting metabolic rate is approximately proportional to $M^{0\cdot7}$, where M is the mass of the animal.[204]

In activity, and especially in flight, the metabolism is raised considerably; in humming-birds from 6·7 ml O_2 $g^{-1}h^{-1}$ to as much as 45·7 ml O_2 $g^{-1}h^{-1}$.[200] In man, the increased oxygen required in activity is supplied partly by raising the rate of heart-beat (the pulse), partly by a greater difference between the volume of the blood contained in the heart at systole and diastole (the stroke volume), and partly by an increased content of haemoglobin in the blood. The first is immediate, the third long-term, while the stroke volume increases to some extent immediately but much more so with training (so that athletes have a low resting pulse, since when they are inactive they need no more oxygen than anyone else, and as they have a large stroke-volume they can get all they need with fewer beats per second). It is likely that birds are similar. The range of pulse rate in the common nighthawk (*Chordeiles minor*) at a wide range of temperatures was from about 100 to about 300 strokes min^{-1}.[203] The scope for raising it in small birds is limited; under resting conditions the maximum for the black-capped chickadee (*Parus atricapillus*)[278] was 1000 beats min^{-1}, for the black-rumped waxbill (*Estrilda troglodytes*)[205] 1020 beats min^{-1} and for four species of humming-bird of various sizes (*Calypte costae*, *Archilocus alexandri*, *Lampornis clemencii*, and *Eugenes fulgens*)[206] about 1200 beats min^{-1}, all of which are a little over twice the minimum levels and represent the probable maximum frequency of contraction of muscle. Some experiments with flying domestic ducks and great black-backed gulls (*Larus marinus*) fitted with transducers and radio transmitters showed that they doubled their stroke volume but did not increase their pulse rate in flight.[96]

The haemoglobin of fowls[227] is very similar to that of mammals. It has the same molecular weight and gives the same shaped curve for the reaction with oxygen, so that it presumably contains four haem units and takes up four molecules of oxygen. The haemoglobins of some reptiles have only one haem unit while others have several, so that no evolutionary story can be made. Once again it looks as if the same efficient system, in this case haemoglobin of molecular weight about 68,000 with four atoms of iron, has been independently selected in birds and in mammals. A further partial similarity is that while fetal mammals have a different haemoglobin from adults, with the curve shifted to the left, the eggs of fowls and red-winged blackbirds (*Agelaius phoeniceus*) have a haemoglobin with a hyperbolic curve.[228]

The dissociation curve for fowl haemoglobin is slightly to the right of those of mammals, which means that the blood has less affinity for oxygen and dissociates more readily. This fits in with the higher oxygen tension in the lungs. In a number of domestic species and in pigeons the blood leaving the heart is, as in man, almost fully saturated, but the oxygen is at a higher tension. The oxygen capacity of the fowl is lower than in man, that of the pigeon slightly higher. In most of the birds about 50 to 60 per cent of the oxygen is removed in the tissues, twice as much as in mammals.

Active birds must also increase their rate of breathing. The control of this, and of the pulse and stroke volume, is presumably governed by some chemical mechanism, an increased rate being caused either by low oxygen or by high carbon dioxide or by acidity (in practice all three, and especially the last two, go together). In mammals the chemoreceptors in the carotid bodies respond chiefly to low oxygen, while response to high carbon dioxide or acidity is through the direct effect on receptors in the medulla. Carotid bodies similar to those of mammals, though of somewhat different distribution and innervated chiefly by the vagus instead of the glosso-pharyngeal nerve, are present in birds, and presumably function in the same way.[83]

The chemistry of energy production in birds has been little studied. We have seen in Chapter 2 that fat is generally the fuel for flight, and since the respiratory quotient of the house sparrow was found to range from 0·65 to 0·81 with a mean of 0·73 it looks as if fat generally predominates as the substance that is oxidized.[180] The mitochondria of the cardiac muscle of pigeons can oxidize cytochromes with adenosine triphosphate even in the presence of respiratory inhibitors such as cyanide, so that they must have some special properties not possessed by those of mammals.[57]

Birds do not seem to put up much of an oxygen debt when flying. It would clearly be dangerous for them to do so, and their voluntary muscles, like all cardiac muscle, have learnt to do without it.

4.25 The physiology of diving

A special case of shortage of oxygen is given by an air-breathing animal that dives. It might meet the difficulty of being out of contact with the atmosphere in three different ways: it might carry a supply of gaseous air with it, it might use its respiratory pigment to supply more oxygen than usual, or it might live anaerobically. No vertebrate uses the first method, which is well-known in insects, except that in various forms it is the device used by human divers. The others are used very widely, in mammals and reptiles as well as birds, and must have been evolved many times.

Few birds submerge for long. Usual times for the cormorant (*Phalacrocorax carbo*) are from 20 to 30 seconds, and for the great crested grebe (*Podiceps cristatus*) from 30 to 35 seconds. The intervals between dives may be very short, in cormorants a third and in auks a quarter of the period of the dive. Corresponding to the short time of immersion, birds seldom go deeper than a few metres. Dives of much longer than a minute, and depths of more than about 20 m have been occasionally recorded, but are suspect. Penguins can survive forced submergence of 5 or 6 minutes, and domestic ducks (whose immediate ancestor, *Anas platyrhynchos*, seldom dives) three times as much, but they are presumably not as active as a feeding bird would be.[313]

The first physiological adaptation is an automatic cessation of breathing (apnoea) when the head is under water, which has been investigated chiefly in domestic ducks. It is in part a postural reflex, presumably mediated by the semicircular canals, and it is said that a duck can be suffocated by holding it with its head downwards. There is also a thermal reflex,[288] breathing being stopped by contact of the naris with water below about 30°C, while high carbon dioxide, which in most birds causes an increased rate of breathing, also induces apnoea. In penguins and domestic ducks the rate of heartbeat falls during the dive (bradycardia), and there is a reduced supply of blood to the pectoral muscles.[313] It seems likely that there is also a fall in body temperature (which has been shown to occur in seals). The pectoral muscles have a high supply of myoglobin, and the oxygen of this is used before anaerobiosis begins. There is, as usual, some build-up of oxygen debt, but it is not as large as might be expected; perhaps birds' muscles, which are mainly used in flying, for which the mechanism of oxygen debt is unsuitable, have lost or have not acquired the full ability so well known in mammals and frogs.

The properties of avian haemoglobin, described above, will facilitate diving. Ducks and loons have a higher concentration of haemoglobin in their blood than game-birds, and diving ducks have a greater blood volume (but not a greater concentration of haemoglobin) than dabbling ducks.[130, 227]

4.3 TEMPERATURE CONTROL

The body temperature of many birds, ranging in size from tits to ostriches, is above 40°C, compared with 37°C which is taken as the average for man.[376, 378] Some species are somewhat cooler than 40°C, but it seems that in general birds maintain slightly higher temperatures than mammals. The two classes must have evolved their homoiothermy independently, but the methods they use are remarkably similar.

It is a condition of good temperature control that there shall be an insulating layer near the surface of the body, both to reduce the metabolic cost of maintaining the body above the ambient temperature and to withstand sudden fluctuations in the surroundings. Such insulation is supplied in mammals by the fur and in birds by the feathers. Birds are better than mammals in altering the disposition of the insulation according to the circumstances. In cold weather they fluff out their feathers, so increasing the amount of air that they enclose, and some of them do this in full sun also; this prevents the radiant heat from reaching the body, while the temperature of the surface of the feathers may rise by five centigrade degrees; (the fleece of the sheep acts in the same way, to prevent heat stroke, but the sheep cannot alter the disposition of its wool). Diving birds retain air amongst their feathers and so maintain their insulation, while diving mammals, whose fur with few exceptions becomes wet, have a layer of blubber beneath the skin to act as a further insulator. The parts of the body not covered by feathers may be at a much lower temperature, and many sea-birds and others can stand on ice without melting it and without losing much heat. Since the feet and legs contain practically no muscle their movements are not inhibited by the fall in temperature. Placing the head amongst the feathers of the shoulder ('under the wing'), as many birds do when roosting, may reduce the rate of loss of heat by 12 per cent.

There are, broadly speaking, two ways in which a thermostat can work: by controlling heat loss and by controlling heat production. Mammals and birds use both, but not in the same way.

4.31 Control of loss of heat

Control of loss of heat, or physical temperature control, is in some ways the simpler. It depends almost entirely on altering the rate of evaporation, but there is nothing in birds corresponding to the sweat glands of primates and some other mammals. Water is lost from the lungs, the air-sacs, and the buccal cavity and bronchi, and its evaporation necessarily takes place as the animal breathes.

In many birds—pelicans, gannets, cormorants, frigate birds, herons, owls, pigeons, nightjars and game-birds—exposed to high temperatures the mouth opens and the throat begins fluttering, so that there is a rapid tidal flow of air over the walls of the buccal cavity and pharynx superimposed on the normal breathing.[30] Sometimes one of two closely related species flutters, while the other does not. In some species the rate of fluttering varies, but in others, once fluttering begins the rate is not affected by future changes in temperature, and the muscular contraction appears to be resonant. In the common nighthawk it is about 600 min^{-1}. However, its amplitude and the length of time for which a bout of fluttering continues, increase with ambient temperature. The pulse is about doubled during this gular fluttering, but there is no consistent alteration in the respiratory rate.[203]

In some of the species with gular fluttering, and notably in the fowl, there is also panting, a change of the breathing to short sharp pants, or tachypnoea, which is used also by dogs. Other species, for example finches and sparrows, simply increase their rate of breathing;[308] that of the house-sparrow, for example, rises fairly steadily from 57 breaths min^{-1} at an ambient temperature of 30°C to 160 min^{-1} at 43°C. In this species the water loss per breath is, as near as can be measured, constant, but in others, such as fowls, the cardinal (*Pyrrhuloxia* (*Richmondena*) *cardinalis*) and the poorwill (*Phalaenoptilus nuttalli*), not only does the rate of breathing go up with rising temperature, but the amount of water lost at each breath increases also. There are various ways in which this could happen, but the most likely is by a change in the extent to which the air-sacs are ventilated. The passage of air through the lungs and air-sacs of birds is still not fully known (see Chapter 2), but there is some evidence that while the posterior air-sacs are filled at every breath, there is little replacement of air in the interclavicular and cervical sacs. It may be that these are ventilated only when the bird has an excessive heat load, whether caused by the activity of flight or by a high ambient temperature. The importance of the air-sacs in temperature regulation in the pigeon is shown by the fact that if the abdominal and thoracic sacs are inactivated the water loss is greatly reduced and the bird's temperature may rise.

Panting depends on the temperature of the head, not of the skin or of the body generally, and is stimulated by a centre in the mid-brain.[295]

4.32 Control of production of heat

The general principle of chemical control is that as the ambient temperature falls the katabolism and heat production of the body rise, the exact opposite of what happens in cold-blooded animals. The increased katabolism may take the obvious form of the otherwise useless type of muscular contraction called shivering, and this can be seen in the fowl. In spite of the fact that man obviously eats more in cold weather, and so produces more heat without shivering, many physiologists for long denied that 'non-shivering thermogenesis' occurred in mammals, but it has now been demonstrated in several species. It is said not to occur in birds,[393] but in view of the difficulty that there was in showing its presence in mammals it is still a possibility. The lower the temperature, the more time a bird spends in feeding.

The relationship of heat production (which is conveniently measured by the oxygen consumption, to which it will be proportional if there is no change of substrate) to temperature is not simple.[77] A typical graph is shown in Fig. 4.2. Starting at near the lowest temperature at which the experimental animals will survive, heat production falls more or less linearly with temperature until a certain point, called the lower critical temperature, is reached. Above this is a band called the thermoneutral zone, where the heat production is constant, and then at the end of this, above a point called the upper critical temperature, heat production rapidly rises.

At low temperatures, the chemical control is such as to maintain a constant body temperature, and there is little or no physical regulation. At the lower critical temperature chemical regulation ceases, and throughout the thermoneutral zone the body temperature is controlled solely by the evaporation of water. Above the upper critical temperature the rate of production of heat rapidly catches up the rate of loss, and when this is overtaken the body temperature must rise. In the cardinal the rate of evaporation of water increases fivefold between ambient temperatures of 33°C (the upper critical temperature) and 41°C, but at the latter temperature only a half of the heat produced is being dissipated (Fig. 4.3).

The upper critical temperature depends on the species, but is generally somewhere within the range 33–38°C. The lower critical temperature varies more widely and usually has some obvious relationship to the conditions under which the bird normally lives;[77] it is 37°C in the paradise whydah (*Steganura paradisaea*), which lives in the tropics, and below zero in the Canadian jay (*Perisoreus canadensis*) and other boreal species that

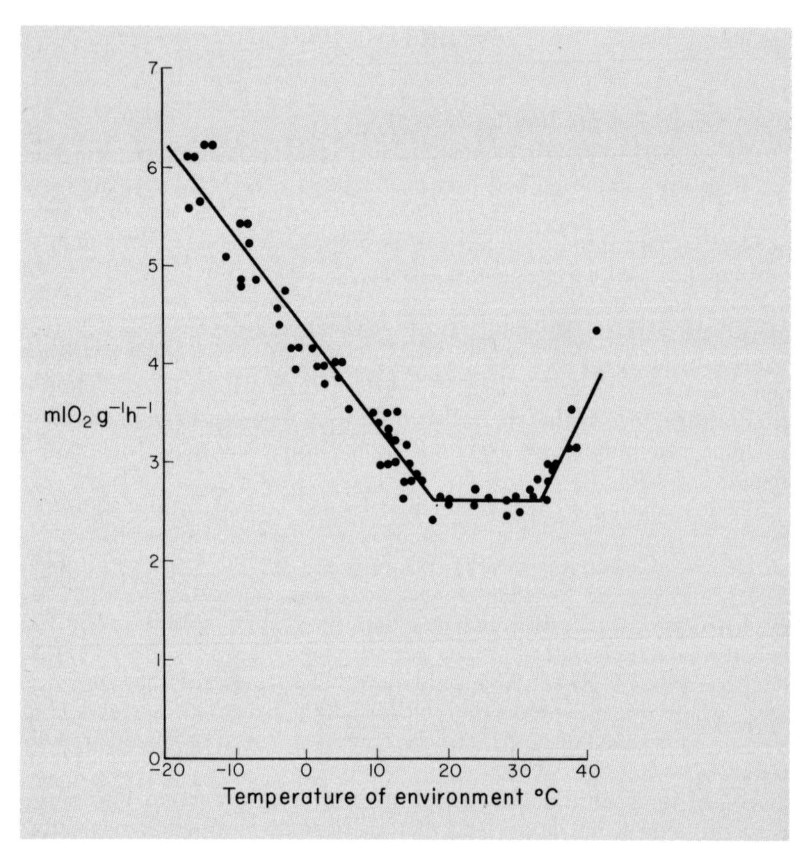

Fig. 4.2 Oxygen consumption of the cardinal, *Richmondena cardinalis,* at various environmental temperatures. Twenty-two birds were used. (From W. B. Yapp, 1970, *Introduction to Animal Physiology,* 3rd edn, Fig. 9.5, Clarendon Press, Oxford. Redrawn from Dawson, 1958, *Physiol. Zoöl.* **31**, 37.)

live under subarctic conditions. It may vary with the season of the year, as in the cardinal, where it is 18°C in winter and 24°C in summer. This species is resident in eastern North America from Ontario to Florida, and so is exposed to a wide range of temperatures. There are some exceptions to this rule of association of lower critical temperature with that of the normal environment; for example the house-sparrow's lower critical temperature is high, at 37°C. This can be partly explained if the sparrow is correctly placed as a member not of the Fringillidae but of the weavers or Ploceidae, most of which live in the tropics. In a few species, such as the common nighthawk, lower and upper critical temperatures are coincident, and there is no thermoneutral zone.[203]

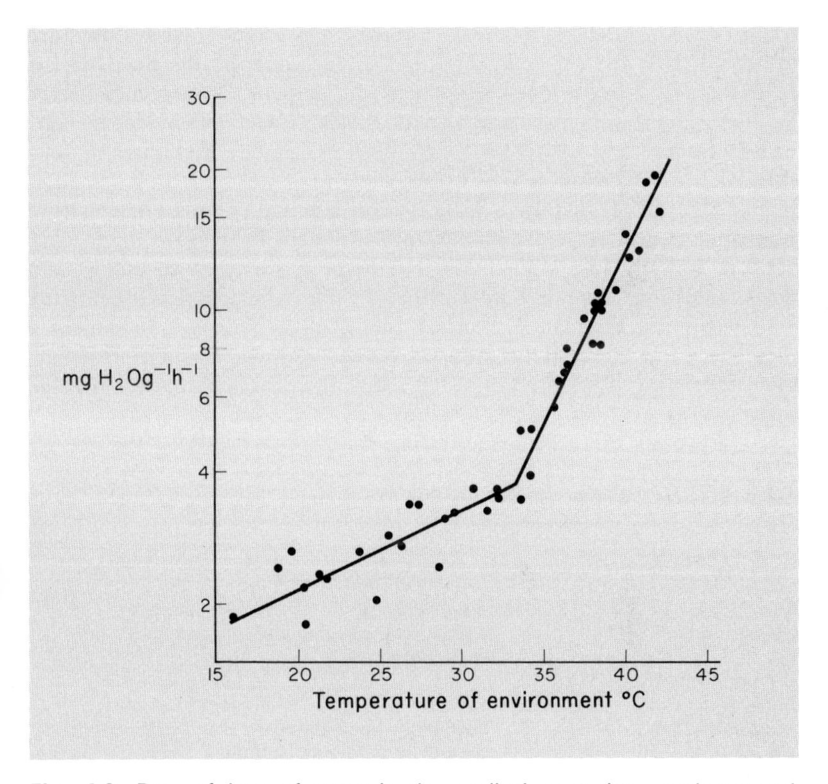

Fig. 4.3 Rate of loss of water in the cardinal at various environmental temperatures. Sixteen birds were used. (From W. B. Yapp, 1970, *Introduction to Animal Physiology*, 3rd edn, Fig. 9.6, Clarendon Press, Oxford. Redrawn from Dawson, 1958, *Physiol. Zoöl.* **31**, 37.)

When at some point above the upper critical temperature regulation has broken down, the body temperature will rise rapidly, and death will ultimately ensue. But the rise in body temperature will itself increase the rate of heat loss by conduction and convection (provided that the ambient temperature is lower), and many small birds allow their temperature to rise a little in hot weather.

The control of shivering in the fowl is both central and peripheral; that is, it can be controlled both by the temperature of a centre in the hypothalamus and by that of the surface of the head.[295]

4·33 Ontogeny of temperature control

Like many new-born mammals, many newly-hatched chicks have no temperature regulation. They are naked and relatively inactive, and are protected from the environmental temperature only by the nest and the parent's body; they are called nidicolous. Temperature control develops gradually and the chemical and physical mechanisms begin independently, usually (in the few species that have been examined) after three or four days. Control is complete well before the young leave the nest. In nidifugous birds (those that leave the nest immediately they hatch) temperature control is present when the young leave the shell; in the fowl the mechanism is working just before this. Nestlings of the western gull (*Larus occidentalis*) are somewhat intermediate,[29] with the temperatures of the youngest chicks fluctuating, but less than does the ambient; the gulls are nidicolous, but the nestlings are well covered with down and quite active. Gular fluttering is fully developed at hatching in the nighthawk[203] (Table 4·1).

Table 4.1 Development of homoiothermy.

Species	Chemical	Physical
Fowl[286]	Before hatching.	Before hatching.
Western gulls[29]	Some present just before hatching, but imperfect.	
Common nighthawk[203]		At hatching.
Sooty tern[170]	Not in chicks.	Present in chicks.
Mourning dove[43]	Both types begin between 3 and 6 days, but are still imperfect at 12 days.	
Vesper sparrows[78]	Imperfect at 4 days, complete at 7–9 days.	
House wren[181]	Control begins day 4, complete at 11–15 days.	
Cactus wren[300]	Not in young nestlings.	In young nestlings.

The brooding parents of some species of nidicolous birds adjust their behaviour to the environmental temperature, and one may expect that this is general.[82, 144, 379] In cold weather they attend the young more, in hot weather less. Similarly an incubating bird may come off the eggs more often in colder weather, but does not stay away for so long, so giving the eggs less chance to cool.

4.34 Torpidity

Many of the old ornithologists believed in hibernation, rather than migration, as the explanation of the disappearance of our summer visitors. None of our regular migrants is known to hibernate, but at least two families of American birds have species that regularly allow their temperature-controlling mechanism to go out of action so that they become torpid.

Humming-birds, small though they are, can maintain their body temperature for a night of eight hours at a wide range of ambient temperatures, even down to 2°C, but at low temperatures they may suddenly cease to regulate and go into torpor, with rigidity and closed eyes.[201, 202] Their oxygen consumption is low and independent of ambient temperatures, so that their body temperature remains about one centigrade degree above that of the surroundings, and they are in effect poikilothermic (Fig. 4.4). This has been observed in several species, but not all individuals behave like this at the same time; torpor comes on most often if the bird's reserves have previously been lowered. After some hours, usually just before dawn, the torpid birds become active and homoiothermic again, even in darkness and at constant temperature. This behaviour will obviously enable them to conserve their reserves.

The poorwill uses its ability to become poikilothermic in a different way.[169] It regularly hibernates, lying up for the winter in rock niches in the Californian deserts. Its temperature is that of its surroundings, and it has no recordable heart beat or breathing. The winter temperature in its habitat is not necessarily very low, at least during the day, being about 20°C, but at a little below this torpidity can be slowly induced in the laboratory, and it can be brought on rapidly by ambient temperatures of 2–6°C. The bird arouses in the laboratory at temperatures above 20°C. The habit may be an adaptation to the absence of moths, on which it feeds, from the deserts in winter.

Torpidity has been described also in a swift, in swallows and in mouse-birds (*Colius*), but if the temperature falls far the birds do not recover; it is unlikely that torpidity occurs naturally or is of any importance in these species.

4.4 NERVOUS SYSTEM AND SENSE ORGANS

Except at a few points, the physiology of the nervous system and sense organs of birds has been little studied. Many experiments have been carried

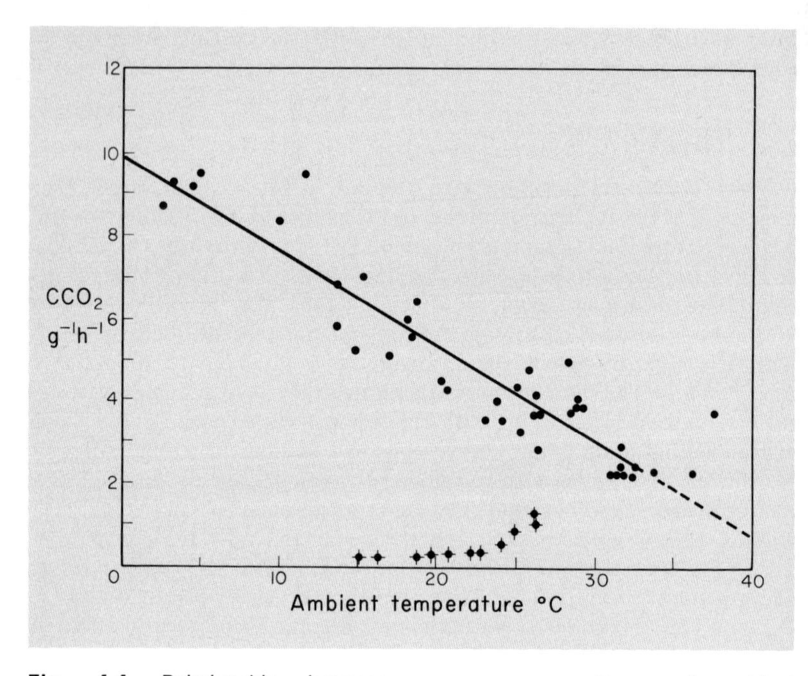

Fig. 4.4 Relationship between oxygen consumption and ambient temperature in homoiothermic and torpid *Lampornis clemenciae*. The regression line fitted through homoiothermic values below thermoneutrality was fitted by method of least squares, and is taken as estimate of thermal conductance. Torpid birds are indicated by a +. (From Lasiewski and Lasiewski, 1967, Fig. 1, *Auk* **84**, 34.)

out on frogs and mammals, and a few on other vertebrates, and it is clear that the same basic principles are used throughout the phylum; many, indeed, apply to invertebrates as well. There is no reason to think that the transmission of the nervous impulse, the action at a synapse, the initiation of a nervous impulse in a receptor cell, or the action of an efferent nerve-ending on its effector, differ in birds in any important respect from those in mammals or other vertebrates. The special features of birds are quantitative, rather than qualitative. In particular, birds have, in several features, achieved a sensitivity and a refinement equalled elsewhere only by the mammals.

4.41 Nervous system

In the nervous system itself birds resemble mammals, and differ from other vertebrates, in two main respects. Firstly, they have a well-developed autonomic system, in which most viscera have dual control by fibres from the sympathetic and parasympathetic systems.[104] The oesophagus and crop of the fowl, for example, respond to parasympathetic stimulation and to drugs in essentially the same way as the visceral muscles of mammals.

Secondly, there is a great enlargement of the brain, which has already been mentioned, in connection with flight, in Chapter 2. Figure 4.5 shows

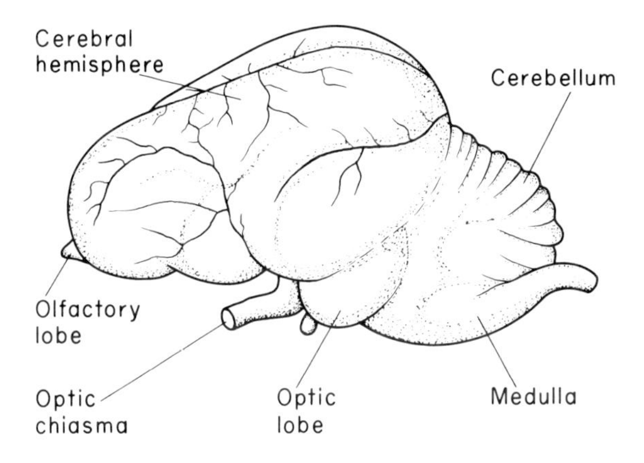

Fig. 4.5 Brain of a parrot. (From Kalischer, 1905, *Abh. preuss. Akad. Wiss. Anhang. Abt.* **4**, 1.)

the brain of a parrot, and may be compared with that of a lemur in Fig. 4.6. It is obvious that while the cerebral hemispheres are big in both, the cerebellum of the parrot is relatively larger. In the fowl and the pigeon, the two species whose brains are usually figured, the cerebral hemispheres are not so large, and do not cover the mid-brain, but they are comparable in relative size to those of such a mammal as the rabbit. The optic lobes of birds are larger than those of mammals, and so are pushed to one side, rather than covered, as the cerebral hemispheres enlarge. The olfactory lobes are small.

Much less is known of the detailed function of the parts of the brain in birds than in mammals, but basically the distribution of functions is the same. The cerebral hemispheres are concerned with the formation of conditioned reflexes, but they seem not to be the only site where these are formed. So far as is known there is nothing comparable to the topographical localization of centres for the special senses that is well-known in mammals. The hemispheres have few striations, but in the absence of knowledge of the detailed functions of the hemispheres, no useful comparison can be made with mammals. The mid-brain, and particularly the

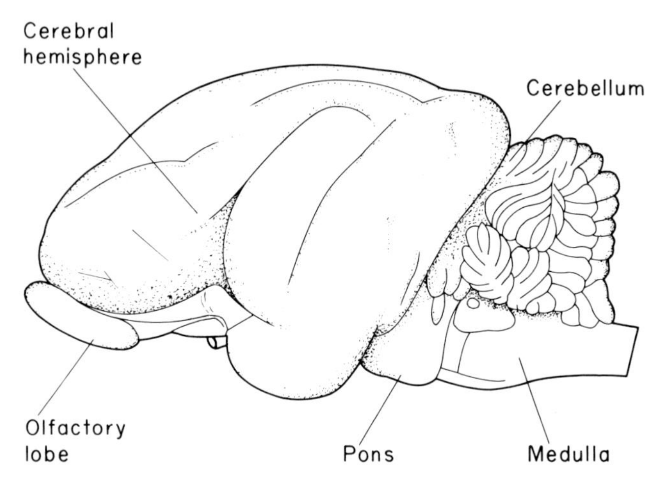

Fig. 4.6 Brain of a lower primate, *Lemur fulvus,* left lateral aspect. (From G. Elliot Smith, 1903, *Trans. Linn. Soc.* (2) **8**, 319.)

thalamencephalon, has similar functions to those of mammals, and the cerebellum is concerned largely with balance.

4.42 Simple sense organs

Even in man our knowledge of the simple senses of touch, temperature and pain is very incomplete, and it is not possible to connect the known senses with various types of receptor organ in the skin and elsewhere. In particular, it seems that the same receptor must not only respond to more than one type of stimulus, but must send impulses to the brain in such a way that the stimuli can be distinguished, a conclusion which is contrary to what was for long part of the folk-lore of sensory physiology. There is, however, no physical impossibility in such discrimination, for just as intensity seems often to be recognized by the frequency of nervous impulses arriving in the brain, so quality might be recognized by the pattern of such impulses.[136] There is some evidence from toads and monkeys that different types of pressure on the skin, which in man would cause different sensations of touch, do cause different patterns of impulse.

Several types of sensory corpuscle have been described in various parts of birds—on the skin, the tongue and the beak, as well as internally.[294] Like the better-known Pacinian corpuscles of mammals, their common feature is a nerve fibre the non-myelinated end of which is surrounded by a number of non-nervous cells, which in the large corpuscles are arranged in layers, like the skins of an onion, to form a lemon-shaped structure. Two examples are the corpuscles of Herbst, and the smaller corpuscles of Grandry (Fig. 4.7), which are especially densely distributed on the soft

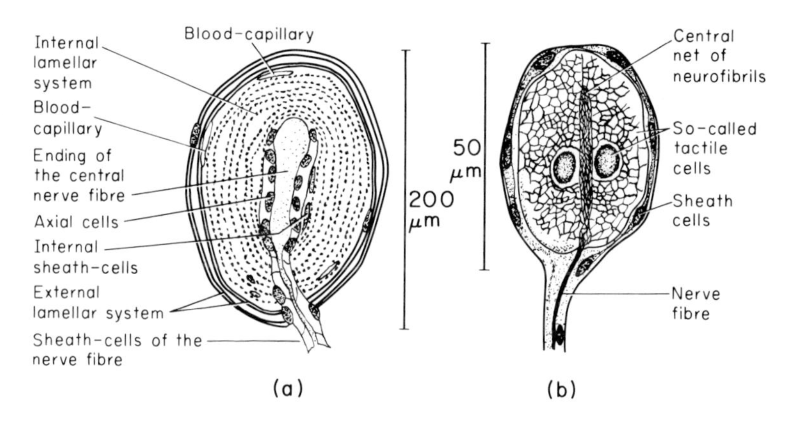

Fig. 4.7 (**a**) Herbst corpuscle and (**b**) Grandry corpuscle, from a duck's bill, both in longitudinal section, not to scale. (Redrawn from Portmann, in Marshall (ed), 1960–61, *Biology and Comparative Physiology of Birds*, Academic Press, London (**a**) after Clara (**b**) after Boeke.)

beaks of ducks and waders. These birds probe the mud, and presumably in this way find their food. The fibres of these corpuscles in the beak are part of the ophthalmic branch of the trigeminal nerve. By analogy with mammals one would expect them to respond to more than one type of stimulus. Some Herbst corpuscles have been shown to respond to vibrations of about 20 to 1000 Hz and may therefore be used to detect movement; for example, corpuscles in the beak might detect movement of an animal in the mud, and those in the leg that of the branch on which the bird is sitting.

4.43 Chemical senses

4.431 Taste

Taste-buds almost exactly similar to those of mammals have been described in birds; in the pigeon they are present chiefly on the dorsal surface of the tongue, but there are fewer than in mammals. It is obvious that some birds, such as tits, can distinguish different sorts of food either by taste or smell (the two cannot be separated except by rigid experiments), since they will take one sort of food and not another irrespective of their appearance, but it seems likely that birds such as fowls and pigeons, which feed largely on seeds and swallow them whole, make little use of taste. Nevertheless geese show preferences for one sort of grain over another,[224] and pheasants can select calcium-containing grit; hen birds increase their intake of this during egg-laying by 50 per cent, while that of the cocks does not change.[192]

4.432 Smell

Physiologists have long maintained that birds have no sense of smell, an opinion based on the small size of the olfactory lobes and on some experiments on conditioning in domestic birds which failed to show any use of odours in this field. Against this wildfowlers have asserted that it is necessary to stand downwind of ducks and geese if they are not to become aware of man's presence, and nearly a century and a half ago Waterton maintained, as a result of experiments, that vultures find their food by smell, not sight.[382]

Although always small when compared with those of most mammals, the olfactory lobes of birds do in fact show a considerable range of size, and similarly there are variations in the size of the olfactory nerves and in the complexity of the nasal chamber; moreover, there is what looks histologically like a perfectly good olfactory epithelium, although its area is small. While smell would appear to be an unimportant sense in most birds, it seems unlikely that it is completely lacking. In man also the olfactory lobes are small, and his sense of smell is negligible compared with that of a dog; but it is not absent.

Olfactory lobes and nerves[25] are relatively well-developed in emus, Procellariiformes, Anseres, Charadrii, the turkey vulture (*Cathartes aura*) and the oilbird, and these groups include some of the birds in which smell seems to be most used. Shearwaters and albatrosses (Procellariiformes) are said to be able to find liver thrown out from ships through thick fog.[24] Especially good olfactory apparatus is found in the kiwi, a bird which is nocturnal, has poor eyes, and can follow trails like a dog.[302, 391]

A few experiments have been claimed to show that pigeons and other birds can form conditioned reflexes to strong smells,[250] but few individuals have been used and some of the conclusions have been criticized.[54] Direct electrical recording from a few fibres of the olfactory nerve has shown that various odorants can initiate impulses in a wide range of species, including not only ducks, geese and vultures, but three species of passerine. There were no great differences between any of the species tested.[371]

The best conclusion from all this is probably that the sense of smell in birds is comparable to that of man. It is present, but in most species is little used.[340] The difficulty in forming conditioned reflexes to olfactory stimuli is probably due to lack of central connections (or, in human terms, lack of interest), rather than inability to perceive the stimulus. Turkey vultures offered a choice of pans containing leaves and pans with leaves in which food was buried, mostly, but not always, investigated the latter first.[282]

4·44 Sight

In gross structure the eye of a bird does not differ much from that of man[292] (Fig. 4.8). It is generally relatively larger, and in some species the

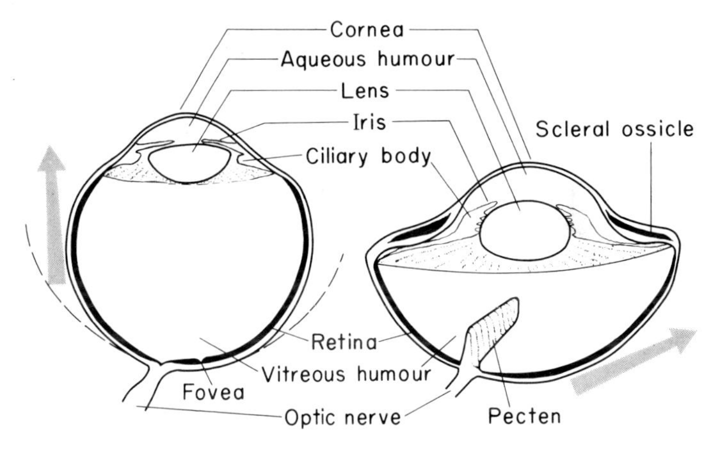

Fig. 4.8 Lower halves of the right eye of a man and left eye of a swan. Position of image-plane of human eye dotted. The arrows point forward. (From Pumphrey, 1948, Fig. 2, *Ibis* **90**, 174.)

two eyeballs almost touch, the orbital region of the skull being constricted and its septum very thin or fenestrated. Perhaps because of this the extrinsic eyeball muscles are small, and birds move their eyes little or not at all; instead, they turn their heads, which they can easily do because of their long and flexible necks.

The eye is not nearly so spherical as man's, generally the front is flattened, as in the figure, so that the sub-spherical lens projects into the hemispherical posterior chamber; in some species, especially those such as hawks that have good vision, the lens is projected forward in a short tube, at the front of which is a nearly hemispherical cornea. The effect of these shapes is that there is very little spherical aberration. Accommodation for near vision is produced by the contraction of striped ciliary muscles which squeeze and thicken the lens; this is the same mechanism as is used by most reptiles. Maximum accommodation in a young man is about 10 diopters; that of birds is about twice this, and in the cormorant, which needs great accommodation to see under water, it is 40–50 diopters.[357]

Growing into the vitreous humour from the optic nerve near the blind spot is a structure, the pecten, which consists of blood vessels in a slight supporting matrix, and is a development of a smaller structure found in reptiles. The retina of birds has no blood vessels, and it therefore seems likely that the primary function of the pecten is to supply oxygen and food to the cells of the retina by diffusion through the vitreous humor. It must, however, cast a shadow on the retina, and it may be helpful in judging movement. Its shape bears some relationship to the habits of the bird. In most species its shadow is a triangle with wavy edges, in the nocturnal owls and nightjars it is a simple bar, while in hawks it is palmate with several radiating fingers.

The light-sensitive cells of the eye are the retinal elements, the rods and cones of human vision. It is usually said that cones are concerned with colour vision and are present chiefly in diurnal animals while rods are more sensitive to faint light but cannot distinguish colours, and are characteristic of nocturnal animals. Electron microscopy has shown that there is no essential difference between them, the shape is unlikely to be of great importance, and wider knowledge of their distribution in mammals has shown that it cannot be simply correlated with habits. The real distinctions are between elements that respond differentially to different wavelengths and those that do not, and between parts of the eye where many elements are connected to a single optic nerve fibre and those where only one or a few elements are so connected. Neither of these has a necessary or regular connection with the shape of the receptor cells.

Over most of the retina of man several elements, perhaps as many as a hundred, converge on to a single optic nerve fibre, but in one part, the area or macula, only four or five do so, and in the pit or fovea in the centre of this there is one-to-one correspondence. Consequently, over most of the retina the image is blurred; we can see an object clearly only by looking straight at it so that the image is on the macula. At the fovea the ability to discriminate between two spots close together is about that calculated on the assumption that for this to be possible each spot must form an image on a different element. In some birds the structure of the retina is such that discrimination over the general surface should be almost as good as that in the macula of man.[87] For such good discrimination to be useful there must also be improvement in the sharpness of the image, and as we have seen the shape of the eyeball is such as to achieve this by eliminating spherical aberration.

Most birds have an area and a fovea, and in this the discrimination may be two or three times the best that man can do. This fovea is in a similar position to that of man, and since the eyes of birds are in the sides of their heads it can be used only in monocular vision; a robin (*Erithacus rubecula*) looking sideways at a worm before pecking at it, is presumably focusing on the fovea. In many birds, such as hawks, swallows, shrikes, humming-birds, kingfishers and terns, there is another fovea in the temporal region, so placed that an object straight ahead of the bird forms an image on the fovea of each eye, and there is good binocular vision, though over a limited angle. The birds which have this feed on moving prey or otherwise need a good appreciation of the exact position of an object straight ahead. Experiments on conditioning (which have been carried out on very few species) have not demonstrated any ability to discriminate patterns of lines better than man can, and some birds do not do so well. The special point about the bird's vision is not that the discrimination is better than man's, but that it is very good over a very wide angle. A bird can see clearly over a large field, whereas man has to turn his eyes or his head and can cover the field only by scanning. So far as they have been studied, reptiles, and mammals other than primates, have much worse acuity.

In some birds there is a third type of area, extending horizontally round the eye. This is found especially in some sea-birds and birds of wide open spaces such as waders and ostriches. It would give good acuity in the horizontal plane, but may also be important as a reference line in balance. The pit-like fovea also probably helps in fixation on an object, since any movement of the image up the sides of the pit will change its distance from the lens and so its size, and make the movement readily detectable. Certainly lizards, which have deep foveae, fixate very steadily on a moving object.

The vertebrates with colour vision are birds, primates, lizards, tortoises, urodeles and teleosts, although in all of these, except (so far as we know) the birds, there are many colour-blind species. In all these groups except tortoises many species have some parts of the body highly coloured, and display them in a more or less elaborate courtship. Colour vision in vertebrates therefore appears to be strongly associated with colour used for sexual purposes.

The colour vision of birds depends, like that of mammals, on retinal pigments derived from vitamin A, but little is known of the details. The spectrum as seen by birds seems to be similar to that of man.

4·45 Hearing

It would obviously be useless for birds to sing if they could not hear each other, and there is plenty of evidence, both from general observation and from laboratory experiments, that the hearing of birds is good, and comparable to that of mammals.

All vertebrate ears are built on the same plan, but those of mammals are peculiar in some respects. There is an external conch or pinna, which at least in some species acts as an ear trumpet to concentrate the sound waves; there are three bones, instead of one, to convey sound across the middle ear; and there is a helical cochlea, which is a differential analyser for pitch. Birds have no pinna (though in owls there is a comparable flap of skin on the anterior margin of the opening), and in most species the ear is more or less completely covered with feathers. There is only the single bone, the columella auris, in the middle ear, but there is a structure very similar in some respects to the mammalian cochlea; it is called by the same name, but it is not spirally wound and is certainly not homologous with the cochlea of mammals, since there is no possibility of its presence in any cotylosaur, the nearest common ancestor of the two groups.

In spite of the lack of homology, the cochlea of birds has the same basic structure as that of mammals (Fig. 4.9). It is divided horizon-

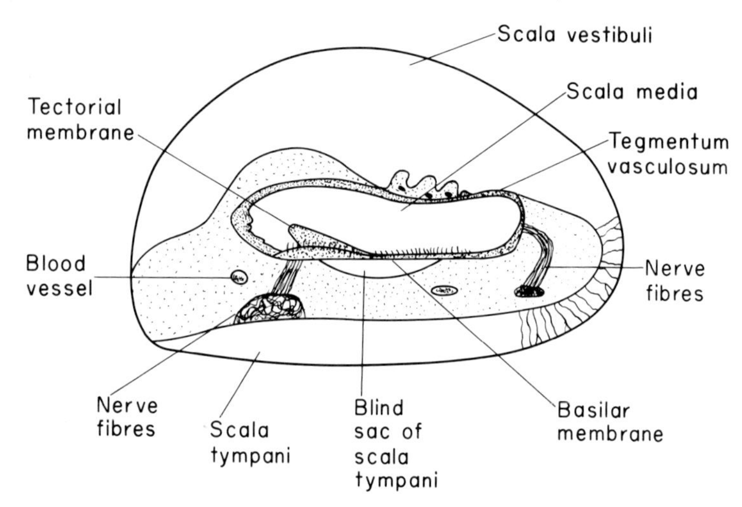

Fig. 4.9 Transverse section of the cochlea of a pigeon, ×c50. (Based on Retzius.[299])

tally into three compartments, of which the upper, the scala vestibuli, does not contain fluid, as it does in mammals, but is filled by a folded membrane, the tegumentum.[315] The middle one, or scala media, bears the sensitive part of the ear, the organ of Corti. The cells of this bend under changes of pressure, caused by the sound-waves and transmitted to the upper and lower chambers, and so initiate impulses in the fibres of the auditory nerve. Corti's organ, like that of mammals, acts as a resonator, different parts of it responding maximally to different wavelengths. The cochlea of birds is shorter than that of mammals, and it seems that in general the range of pitch over which birds can discriminate is a few octaves less than in man.[44] Song-birds can distinguish notes from a few hundred Hertz to about 10,000 or a little more. Most of the few non-passerines studied have an upper limit of between 5000 and 7500 Hz. Over this range their discrimination is as good as man's. The tegumentum, through which the sound-waves must travel before they reach the basilar membrane, must damp them, and this should mean that the ear can respond to sounds of shorter duration than can be appreciated by the ear of mammals. There is some evidence that this is so, but the claim that the sound spectrogram (see Chapter 7) of the song of the chaffinch shows short notes inaudible to the human ear is false, since they are readily detectable to anyone accustomed to listen to music.

The chief function of hearing in birds is in communication. With very few exceptions, such as the ostrich, even the hoarsest of birds has a range of voice far wider than that of any mammals except some primates; even the raven (*Corvus corax*) is not restricted to the traditional croak. To match this variety of sounds they need good ears and they have them. In the song birds, voice has been developed to a very high degree, and is part of the sexual life. That birds normally of more restricted voice have better ears than they need is suggested by the ease with which such birds as jackdaws (*Corvus monedula*) and the grey parrot (*Psittacus erithacus*) can be taught to talk. There is a problem in evolution here.

There is abundant anatomical evidence that the ears of owls are better than those of other birds.[292, 315] They have a much longer basilar membrane, and the ears have an asymmetry which would be expected to have a function. There is electrical evidence that they are sensitive up to about 20 kHz. There is now little doubt that the owls use their good hearing for finding their prey, both from the rustlings that the mice make in moving through the grass and from the squeaks that they emit. It has been shown that the asymmetry of the ears should help in the estimation of the location of the point of origin of high-pitched sounds, which in man is notoriously difficult.

5

The Endocrine Control of Reproduction

The seasonal rhythm of reproduction in birds has been known from time immemorial, but it was not until the 1920s that any serious or successful attempt to analyse its causes was made. Rowan[306] then showed that in the slate-colored junco (*Junco hyemalis*), a migratory finch of the Canadian forests, the most important factor in producing breeding condition was light. Since then some thirty species of several families, mostly North American passerines, have been investigated, and the pituitary and the hypothalamus have been added as links in the chain of causation. In spite of all this work, the story is not yet complete, and different schools of workers interpret the results of experiments somewhat differently.

5.1 THE FACTS OF REPRODUCTION

5.11 The gonads[32]

The reproductive organs of birds are built on the usual vertebrate plan, which is fundamentally hermaphrodite. At its first appearance the gonad is indeterminate in sex. As it grows, if its inner part, or medulla, becomes prominent, it forms a testis, while if the outer part, or cortex, prevails it forms an ovary. Unlike most vertebrates the male is homogametic, so that the hormonal control of the development of the glands, ducts and other structures appropriate to one sex or the other must be slightly different from that in other classes. Some limited and difficult experiments on ducks, and on a few amphibians and mammals, suggest that this is so. In the duck a penis and the male type of syrinx develop in a genetic male or in castrates of either sex, but they are suppressed in the presence of ovarian tissue. The male therefore appears to be the neutral form, and the female to be a suppressed male, whereas mammals are the other way round. However, the case is not simple. In ducks, as in mammals, the Müllerian duct (oviduct) persists unless it is repressed by the presence of a testis.

Whatever the exact mechanism, the balance is delicate. Old domestic hens often develop some of the secondary characters of cocks, such as the comb, and there are a few well-authenticated records of complete sex-reversal, with former hens becoming potent cocks. Intersexes, in which the characters of male and female are mixed in various ways, are well-known in many vertebrates, and some of the occasional field records of singing female birds may well be based on such.

In most birds the right ovary and oviduct do not develop, but remain as rudiments; the sex reversal of fowls depends on the suppression of the production of hormones in the functional left ovary by old age or disease, so that the right gonadal rudiment develops; the part that grows is the medulla, so that it becomes not an ovary but a testis, with consequent readjustment of the ducts. In some of the Falconiformes, the right ovary (but not the right oviduct) is large, but it is rarely functional.

In both sexes the gonads undergo a very marked seasonal variation in size. In winter they shrivel almost to nothing, and in spring they increase as much as 500-fold.

Although birds have internal fertilization most do not have a penis. In ducks and geese one is present which consists of two incompletely joined halves; it protrudes from the vent only when the male is excited. A penis is present also in the ratites and a few others. In the game-birds and some passerines, such as finches, there is a papilla on the wall of the male cloaca, which is inserted into the female in copulation.[403]

In poultry the sperms are normally stored after copulation in a pouch near the lower end of the oviduct (at the junction of the parts miscalled uterus and vagina) and are released in gobbets about the time of ovulation, after which they ascend the oviduct for fertilization in its upper part.[219] The turkey (*Meleagris gallopavo*) can lay fertile eggs 30 days after treading, and the fowl after 20 days.[150]

A few birds, such as domestic ducks and hens and the African grey parrot, ovulate spontaneously, but most need the presence of a male. Foreplay, a series of behaviour patterns in which both sexes take part, may then occur and lead to copulation. The connection of this to ovulation (the release of an egg from the ovary, not the laying of an egg) seems not to have been investigated. Domestic pigeons do not lay eggs if they are alone, but may do so, and if so must have ovulated, if two hens are confined together. One may even do so if she is caged alone but provided with a mirror, so that she can display to her own image.

Most birds are sexually mature in the spring after that in which they were hatched, that is when they are a little less than a year old, but in some, especially the large ones, there is a longer period of adolescence. Rooks breed first at two years, many gulls at three or four years, the fulmar probably at seven, eight or nine years, and the royal albatross (*Diomedea epomophora*) probably at eight years.[316] Some tropical species such as the budgerigar and the Australian zebra finch (*Taeniopygia castanotis*) can breed at a few months old.[171a]

5.12 Breeding seasons

Copulation, fertilization, laying, the brooding of the young, and their growth to a stage when they can look after themselves, follow rapidly one after the other, so that the whole activity of production is crowded into a few weeks, or at most, months; there is nothing comparable to the temporal separation of the association of the sexes from the birth and suckling of the young that occurs in nearly all the larger mammals and many of the smaller ones. The resulting breeding season, when the gonads of both sexes are large and active and all the other features of reproduction can be seen, is, in temperate climates, usually the spring or early summer, the average time for a species being later the nearer it lives to the pole. There are only two sorts of exception to this in the northern hemisphere. A few species nest very early in the year; for example the crossbill may breed in February in Scotland and in January in North America, though it is also found nesting much later than this; the raven regularly nests in February or early March. A few other species have a very extended season, so that nests of different individuals may be found in many months of the year. The mourning dove (*Zenaidura macroura*) nests in the United States in any month except November and December, but chiefly from March to September,[353] and the wood-pigeon in southern England nests in every month.

In south temperate regions, such as New Zealand, some species of sea-birds—petrels, gulls and terns—breed in the autumn, in places where other species of the same groups breed in the spring.[107] In birds living in deserts in Western Australia there is much breeding in abnormally wet autumns; this is most marked in the driest regions.[317]

Most species breed only once in the year, but a few regularly have two broods, and in most a second nest may be made and a second clutch of eggs laid if the first is for any reason unsuccessful. There are a few examples of a migrant breeding twice in two different places. Redpolls (*Carduelis flammea*) may breed early in the year in spruce forests in southern Scandinavia, and again later in birchwoods in the north.[285] Less certainly the quail (*Coturnix coturnix*) may breed twice in different parts of France.

A few large birds, such as condors (*Vultur* and *Gymnogyps*), albatrosses and penguins, take more than one year to rear their young, and so do not breed every year. In the arctic many individuals do not breed, apparently because the external factors that should lead to the development of the gonads are not adequate.[234]

It is a truism that, if the species is to survive, the young must be produced when there is enough for them to eat, but it is a big jump from this to assume, as is often done, that natural selection has ensured that they are produced when there is most for them to eat. Other conditions must be suitable as well, and the environmental conditions such as high temperature, that tend to make food abundant, are likely to promote sexual maturity in birds even in the absence of any selection. The supply of food for most temperate birds is, nevertheless, near its maximum at the time when they breed, and the birds that nest exceptionally early or throughout the year are those that feed on vegetable matter, or on animals that are plentiful at all times.

In the tropics, the terms 'summer' and 'spring' are meaningless, and indeed are not part of the language. Instead, there are rainy seasons and dry seasons, sometimes one of each in a year, sometimes two. The descriptions are, however, relative, and in the regions dominated by the evergreen tropical rain-forest rain may occur in any month and on any day. Further from the equator the dry season may be really dry, with no rain at all for two or three months. There may also be a cycle of hot and cold seasons, which is largely independent of the rainfall. Even in the rainy season there is usually much sunshine, and except in the areas where there are prolonged periods without rain vegetable growth is active throughout the year, and insect life is correspondingly abundant. The variation in length of day is small and slow.

Under these conditions it would be surprising if seasonal breeding followed the same regular pattern as it does in temperate regions and there is abundant evidence that it does not. Our knowledge of it is still very incomplete; ornithologists are few, and finding nests in tropical forest, or even in savannah, is difficult, so that it is not surprising that for species after species the handbooks say 'breeding unknown'. Most of the published information on breeding seasons deals either with large conspicuous birds such as raptors, or with birds that nest in gardens. There is some evidence from England that woodland birds that have taken to nesting in gardens also alter some of their other habits and tend to breed earlier, so that information obtained from such birds must be treated carefully.

No general conclusions can justifiably be drawn from the information available. In Guyana (British Guiana) the peak month for nests is May, the wettest month of the year;[74] in the Republic of Ecuador, about the same distance from the equator but on the other side of the continent, the peak is from February to April, after the end of the occasional rains;[229] at 1° south, in Brazil, it is in October, towards the end of the dry season.[289]

In east Africa some species show two peaks, and probably some individuals breed twice in the year.

On some oceanic islands, with a uniform climate and a constant food supply, some sea-birds have altogether abandoned the yearly cycle. On Ascension, for example, the sooty tern (*Sterna fuscata*) begins to breed a little earlier each year, so that it produces about four clutches in three years; the mean interval over 12 years was 9·6 months.[59] The yellow-billed tropic bird (*Phaeton lepturus*) and the red-billed tropic bird (*P. aethereus*) on the same island breed at intervals of 5–10 and 9–12 months respectively, the variation being caused by the failure or success of the previous breeding cycle.[347]

The facts that have to be explained are, in summary, that nearly all the temperate birds have a breeding cycle which follows that of the sun, activity being at its maximum when length of day, intensity of light, and temperature are all increasing, but before any of these has reached its peak; that there are exceptions to this rule; and that in the tropics, where two of the three factors that make the seasons in temperate regions do not vary in a regular way and the third, length of day, changes very little, most birds nevertheless still have seasonal breeding.

5.13 Hormones connected with reproduction

5.131 The adenohypophysis[28]

The adenohypophysis (part of which was formerly called the anterior pituitary) of mammals produces three hormones which have various effects on the reproductive system and are collectively known as gonadotropins. It seems likely that closely-related substances are present in all vertebrates, but, to judge from other better-known hormones such as those of the neurohypophysis, it is probable that there is a family of related substances which may be included under each name. The analysis of the gonadotropins produced by birds has hardly begun, and most of our knowledge comes from experiments on the injection of mammalian hormones, which are readily available.

The action of these hormones in mammals is complicated, and there are some specific differences, but briefly, follicle-stimulating hormone (FSH) causes ripening of the ovarian follicles and induces the ovary to form oestrogens, while luteinizing or interstitial-cell-stimulating hormone (ICSH) causes the transformation of the follicle, after ovulation, into the corpus luteum. Oestrogens depress the secretion of follicle-stimulating hormone, so that there is a system of feedback. Luteotropic hormone or prolactin (LTH) stimulates the secretion of milk by the mammary glands. In the male interstitial-cell-stimulating hormone maintains the secretory activity of the interstitial cells of the testis.

All three of these mammalian hormones have effects when injected into birds, and their analogues probably have natural functions. Both follicle-stimulating hormone and luteinizing hormone will cause enlargement of the testis in a hypophysectomized capon, but luteinizing hormone causes the comb to grow as well.[270] The luteinizing hormone in the blood of a laying hen varies throughout the day, and the peaks appear to have some connection with ovulation. Follicle-stimulating hormone causes the ovaries of canaries (*Serinus canaria*) to develop in any month of the year, but does not always produce all the subsidiary parts of the breeding behaviour.[342] Prolactin[247] has only secondary effects, such as the loss of feathers of the brood patches (Chapter 6) and the development of the pigeon's crop glands. It does not induce broodiness in cocks but does induce some other forms of female behaviour, and it reduces the size of the testes, possibly by shutting off the supply of follicle-stimulating hormone.[269]

5.132 The sex hormones

Both types of sex hormone are produced by the gonads of both sexes, but the female hormones are produced in larger quantity by the female, the male hormones by the male. There is probably only one type of male hormone, or androgen, comparable to the mammalian testosterone, but there are two groups of female hormones: oestrogens, produced by the follicles in the ovary, and progestins (of which the mammalian progesterone is the best known) produced by the interstitial cells. Their effects are varied, but while oestrogen is responsible for some of the more fundamental things, such as the development of the oviducts and the vascularity of the brood patches, progesterone induces sexual behaviour.

5.2 LIGHT AND THE PITUITARY

5.21 Experimental induction of breeding

Full breeding activity includes the swelling of the gonads, the production of active gametes, the behaviour leading to fertilization, and adequate nesting and care of the young. In many species it involves also secondary features, such as migration, song and the development of a special breeding plumage. It is by no means necessary that all these should be produced by the same mechanism, or started by the same external stimulus, but obviously all must fit together in the correct sequence. It is likely that there would be an internal chain of causation, some features of the reproductive cycle being in themselves the initiators of others. Anything caused by a high concentration of hormone, for example, is not likely to occur while the gonads are small in size, for then the interstitial tissue, which secretes the hormones, is as negligible as the germinal epithelium, which produces the eggs or sperm.

Experimental breeding condition can only be said to have been achieved when the whole cycle has been demonstrated following the application of controlled conditions, but useful indications may be given by partial responses. The assumption of breeding plumage, and such things as song, the laying down of premigratory fat and migratory restlessness (which are discussed in the next chapter) have all been observed following various experimental treatments. Most of the work on manipulation of the environment has been done on finches, and much of that concerned with the influence of hormones has been done on domestic poultry.

In a number of species, artificially increasing the length of day, at any time from late autumn onwards, will induce growth of the gonads and some sexual behaviour.[108, 401] In many experiments the development of males has been complete, with production of active sperm and full reproductive behaviour, but the development in females is in general only partial. Some game-birds have been induced to lay eggs out of season by increasing light, but passerines go only as far as some swelling of the ovaries without oogenesis.

Many experiments have been carried out to investigate the parameters of the light ration necessary for stimulation, but many of these have no direct reference to the conditions in nature; no normal bird is subjected to days of four hours and nights of eight hours. Nevertheless these experiments may be useful in suggesting how the light works.

In general, what matters is the number of hours for which the light is above a certain minimum intensity during the 24 hour day. The increase in the length of the light period need not be gradual, as it is in nature. In some experiments a split light ration causes the testes to develop faster than does a single ration of the same total length; for example in the white-crowned sparrow (*Zonotrichia leucophrys*) six hours of light during the 24 hour day, given in equally-spaced periods of 50 minutes, produce the same rate of growth as a continuous period of 12 hours. Fowls can be stimulated to lay more eggs just as well by flashes of 15 seconds duration repeated at 15 minute intervals as by long days.[115] These results can be explained by the hypothesis that during the light period there is rapid production of a substance which is slowly destroyed during the dark period. In another series of experiments rations of eight hours of light and 16 hours of darkness failed to produce a response, but rapid alternations of one hour light + two hours darkness, which gave the same total of each, caused the gonads to develop. In other experiments, on the slate-colored junco, a different type of result was obtained; eight hours of light + 16 hours of darkness were ineffective, 16 hours of light + 32 hours of darkness were effective.[402] All these results agree in suggesting that during the light period something important happens that brings the birds into or towards breeding condition. In several experiments red light has been shown to be more effective than other wavelengths.

One would expect that the effect of the light would be mediated through the eyes, and that this is so is shown by the reduced response of birds with cut optic nerves. But such birds do still respond, so that the light must be picked up by some other receptor as well.[32] In the domestic duck the hypothalamus and the rhinencephalon are the sensitive parts; light conducted to the brain along fine quartz fibres acting as wave guides is effective whether it is red, green or blue (Fig. 5.1). In the intact animal much red light, but not green or blue, reaches the brain directly through the orbit, which would account for the greater effectiveness of red.

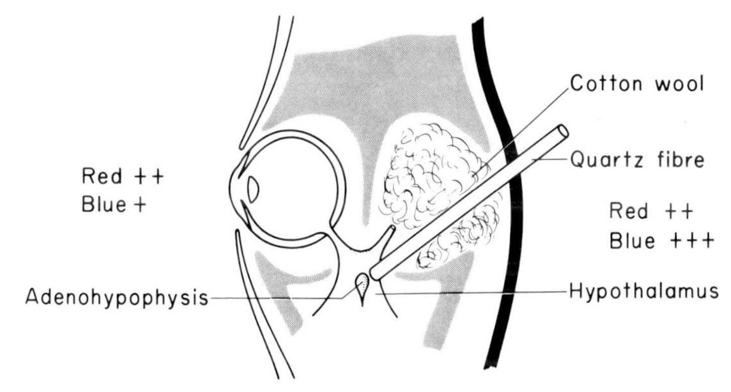

Fig. 5.1 Horizontal section of the head of a duck to illustrate the operation for conveying light direct to the pituitary. Intact eye and orbit on the left, operated ones on the right. Shaded area shows bone. The effects of red and blue light in causing sexual development in each condition are shown. (Redrawn from Bénoit in Grassé, 1950, *Traité de Zoologie* (15), Fig. 341, Masson et Cie, Paris.)

The story can now be summarized and brought into line with what is known for mammals (Fig. 5.2). Light falling on the eyes or directly on the brain initiates nervous impulses which stimulate the hypothalamus to produce neurohumours. They travel along nerve fibres to the median eminence and then, in the blood vessels of the hypophyseal portal system, to the adenohypophysis. This is stimulated to produce gonadotropins, which pass in the blood system to the gonads, which are stimulated to grow and to produce their own oestrogens or androgens.

There is clearly still much to be learnt about this mechanism. The nature of the pituitary hormones, in particular, is still unknown.

Although every species used shows at least some sexual development when it is given increased light out of season, there is for most species a time during the autumn when no amount or intensity of light will produce any response. This is described, but not explained, by saying that after the breeding season there is a refractory period. In the finches that have been used in America it ends in late October or in November or December,[318] but it can be manipulated to some extent by altering the light ration; it can be prolonged if the bird is maintained throughout the

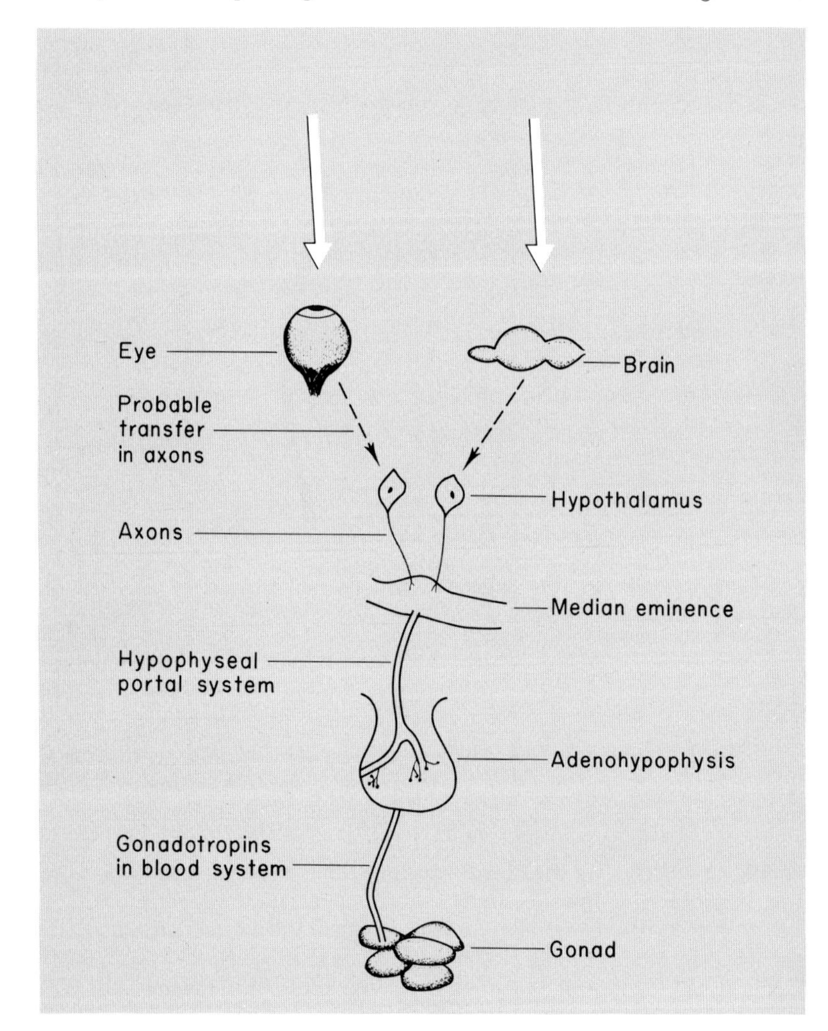

Fig. 5.2 Diagram of pituitary relationships.

autumn in 20 hour light and four hour darkness, or shortened by nine hour light and 15 hour darkness in summer.[400] The wood-pigeon, a bird that can breed in any month of the year, appears not to have a refractory period, or if there is one it is very short.[217] One could explain the refractory period by a hypothesis of negative feedback from the gonads to the pituitary, comparable to what is known to happen in the female mammal, where the oestrogen from the ovary inhibits the production of follicle-stimulating hormone, but there is no experimental confirmation of this. Pituitary activity in the fowl can be suppressed by mechanical stimulation of the oviduct,[171] so possibly the presence of an egg in the oviduct acts in the same way, and prevents the formation of a second egg. This could control the spacing of the eggs, and it is conceivable that it might become adapted to stop ovulation altogether. Whatever the origin of the refractoriness it is at the level of the pituitary or above, since the gonads of refractory birds can be made to develop by injections of gonadotropins. A Lincoln's sparrow (*Melospiza lincolnii*) has been induced to lay eggs by these means.[404]

The refractory period is accompanied by histological changes, especially in the testis, in which the interstitial cells regenerate. Similar changes can be produced by injections of prolactin or by removal of the pituitary.

The autumn sexual behaviour of many species, shown chiefly by song and the premigratory restlessness and deposition of fat (Chapter 6), may indicate the natural end of the refractory period. The length of day is then about the same as that in spring when the birds come into full breeding condition. Some rooks do in fact produce spermatozoa or ovulate in September, although nesting seldom occurs.[236]

5.22 The breeding cycle[109]

All the experiments that have been referred to so far show that the length of day is an important influence on the breeding cycle, but they do not give a full explanation. In England gametogenesis may begin in the robin during the first week in January, when the days are only 14 minutes longer than they were at their shortest in mid-December, and the daily increment is only 1·5 minutes.[235a] The period between sunrise and sunset is a bare eight hours, which is less than that needed to cause the growth of the gonads in most of the experiments, and for gametogenesis to occur at the beginning of the year growth must have begun earlier, on a shorter or diminishing day. Further, the gonads of many migrants begin to develop before they leave their southern home, and many tropical birds have seasonal breeding in an almost constant day-length.

These difficulties seem to have two possible explanations, which are not mutually exclusive. The bird may be influenced by factors other than light, and it may have an internal rhythm on which the external factors have a regulating effect. Tropical birds would seem to be excellent material for investigating these possibilities, but only one, the red-billed dioch (*Quelea quelea*), from the equatorial regions of Africa, has been much used.

It lives in dry, but not desert, country, and breeds at variable times when there is a flush of green vegetation. Birds from east Africa kept in London in constant conditions of a 12 hour day, 22°C and plenty of food, maintained their natural breeding rhythm, in all but a few details, for three seasons.[213] It thus clearly has an internal rhythm. In nature it does not have quite such constant conditions, and first-year birds can be stimulated to breed by providing them with green grass, on the seeds of which they feed, while older birds can be stimulated by rainfall or high humidity, which will in nature cause the production of fresh grass[237] (one may speculate that this is a conditioned reflex). In the laboratory the birds respond also to light, but have a brief refractory period. It seems likely that their breeding seasons are partly innate, depending on an autonomous internal rhythm, and that food or water in some form accelerates gametogenesis, so that, in co-operation with the birds' wandering habits, they enable it to take advantage of good breeding conditions whenever and wherever they occur. The reaction to light is probably vestigial and unused.[84]

There is less good evidence that the short-tailed shearwater (*Puffinus tenuirostris*), which migrates from the Aleutians across the equator to breed in Tasmania, also has an internal rhythm.[238] The response of temperate birds to light is not incompatible with such a rhythm. The cycles of fat deposition and moult in white-crowned sparrows persisted for one year on a constant ration of 20 hour light and four hour dark, while that for fat, but not moult, persisted for the same time on eight hour light and 16 hour dark.[185]

There is evidence, both experimental and ecological, that factors other than light may help in the timing of the cycle, or at least must be such as not to inhibit it.

Some of the birds exposed to long days in winter have come into breeding condition at very low temperatures, some of Rowan's juncoes even at −47°C. Under natural conditions no bird would reproduce, or could survive, at such temperatures. Nesting has been shown to be statistically earlier in warm springs in several species, and sudden falls in temperature may cause birds such as tits to desert their nests, even after eggs have been laid, or, at an earlier stage, to break up their pairs and return to the flock. Conversely, unseasonal warm weather in winter may cause birds to start singing and even occasionally to lay eggs.

Three species of finch kept on a short length of day of nine hours for up to 18 months, but with temperatures fluctuating naturally with the seasons, showed slight gonadal development in spring.[387] Some of the other accompaniments of the cycle were more marked, with specific differences. Juncoes and fox sparrows (*Passerella iliaca*) put on fat and showed migratory restlessness, although rather later than usual, while white-throated sparrows did not show these effects, but underwent a prenuptial moult.

It seems likely that in many tropical birds rainfall in some way determines the time of breeding, as it does in the dioch. In Western Australia also many desert birds have inactive gonads as long as the drought persists, but spermatogenesis and oogenesis begin very soon after rain irrespective of the length of day or its direction of change.[178]

In many species, and especially in the female, complete breeding behaviour only occurs if certain psychological and ecological conditions are satisfied. Oogenesis, as has been said above, usually only occurs if a male is present, and both sexes may need to go through characteristic patterns of display. Eggs of most species are not laid until a nest has been built, and that cannot happen until a suitable site has been found. There is strictly no consummatory act, but a continuous cycle; however, the presence of eggs in the nest, or their contact with the brood patch, inhibits the laying of further eggs in many species; in others, the number laid is fixed within narrow limits.

In the white-crowned sparrow the level of gonadotropins in the pituitary is minimal in autumn, but during the winter it increases, and it triples before there is any detectable increase in the gonads; this suggests that they are manufactured continually, but not released into the blood.[188] By the time the gonads are at their maximum in May the gonadotropins have tripled again. While in males there was no difference between captives and free-living birds, captive females in May had only about one quarter the gonadotropins of wild-caught birds. Presumably other environmental stimuli besides light are necessary to stimulate the female pituitary.

Some of the stimuli of nesting, such as sitting on the eggs, probably cause the adenohypophysis to secrete prolactin which may inhibit the production of gonadotropins and so the laying of more eggs. It certainly, in co-operation with oestrogen, causes broodiness and brood patches (Section 6.23), and probably initiates the regressive phase and the refractory period.

In summary, the bird has a cycle of activity, almost always annual, which may be innate and internal. This cycle involves the hypothalamus, the adenohypophysis, and the gonads; it is most often maintained and controlled chiefly by light. If we start with the refractory phase, it seems that when the bird, by natural regeneration, comes out of this, the light ration may be enough to start some sexual behaviour and even gametogenesis, so that the gonads must be active, but the conditions of winter, whether low temperature or short days, stop all this. Somewhere about midwinter, perhaps by the very small increments of day length, perhaps by occasional days of bright sunlight, perhaps by the effluxion of time, the gonads start growing again; and in spring, under the influence of the longer days on the hypothalamus and adenohypophysis, they grow rapidly. To some extent the different specific patterns may be connected with the different conditions under which the birds live. In many species light has been displaced as the main regulating factor by other things.[216]

5.3 SOME OTHER HORMONAL RELATIONSHIPS

Some of the influences of hormones on aspects of behaviour, such as migration and song, are considered in the next chapter. We may note here influences on the secondary and tertiary sexual characters, and on some other things.

5.31 Plumage

The feathers of birds, as well as being used in temperature control and flight, are the vehicles for the chief colours and often for conspicuous form. They may distinguish not only species, but sexes, ages and time of year. The possible combinations and variations are enormous, and what follows here is a much-simplified account.

There is a minimum of three types of plumage during the life-history; down, juvenile, and adult. Between each one and the next is a moult, in which the old feathers are shed and the new ones rapidly develop. In the downy plumage there are no fully-formed feathers, and the colour is indifferent or cryptic. The juvenile plumage also is usually inconspicuous; it may resemble that of the female, more rarely that of the male, or be similar to that of both sexes where they are alike; sometimes it is unique. The adult plumage is usually acquired within the first year of life, so that the bird wears it when it breeds at about 11 months old; sometimes, as in gulls and gannets, which do not breed until later, the juvenile plumage is retained for three or four or more years. Once they are adult, some birds have only one moult in the year, in the autumn, others have one in the spring as well.

Possibly the majority of species are at least to some extent sexually dimorphic, but in most temperate species the differences are small and hardly recognizable; the female great tit (*Parus major*), for example, has a slightly narrower black band down the breast than the male. Where temperate sexes have obvious differences, as in the chaffinch, they are usually retained throughout the year, but in some species the male's distinctive dress is acquired only at the spring or prenuptial moult, and is lost again when breeding is over. A European example is the ruff (*Philomachus pugnax*), which puts on its characteristic neck-feathers, from which it gets its name, from March onwards. The most extreme examples of seasonal male plumage are found in some tropical birds. The males of most species of the African Ploceidae out of the breeding season are dull and brown and resemble the females, but before nesting they acquire coloured feathers, or tails three times the length of the body (Fig. 5.3). Similarly

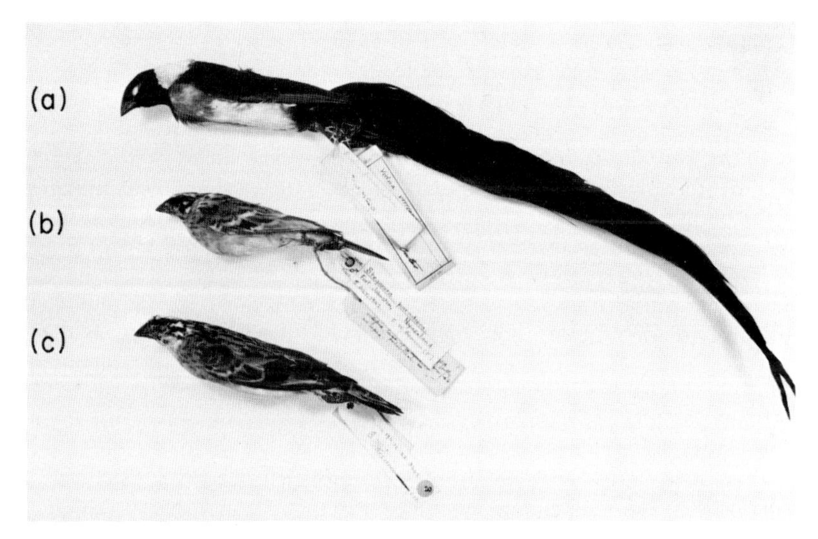

Fig. 5.3 Plumages of the paradise whydah (*Steganura paradisea*): (**a**) male in breeding dress (**b**) male in eclipse (**c**) female, ×0·25.

the males of the sunbirds (Nectariniidae) are dull brown for most of the year, but have brilliant iridescent plumage when breeding.

In a few species both sexes put on a special breeding dress; for example both male and female of the black-headed gull (*Larus ridibundus*) acquire a dark brown head in March or thereabouts, but it becomes white, with some grey, in autumn.

The control of moult is not yet clear. The thyroid is certainly important, for in several species moult does not occur if the thyroid is removed or repressed, and in many administration of thyroxine will induce a moult out of season. During the moult the thyroid is active, and there is often a rise of basal metabolism in line with this.

There is, however, much specific variation. Since in many birds the prenuptial moult (and hence, negatively, the postnuptial one) is obviously connected with the sexual display that precedes or accompanies breeding, it would seem likely that it might be determined by the sex hormones or those of the pituitary, but the evidence is incomplete, and little work has been done except on poultry.[32, 398, 399] In these, castration has no effect on either the down or the juvenile plumage, but in those breeds, such as the Leghorn, in which there are two juvenile plumages, castration of males during the first possibly hastens the moult, while in the female it leads to the assumption of a second juvenile plumage of the male pattern. A precocious change to the adult plumage can be induced in herring-gulls (*Larus argentatus*) by androgens. The adult plumage of pigeons, crows, sparrows and starlings is unaffected by castration or by sex hormones.

The influence of sex hormones on the plumage of dimorphic species falls into two main types.

In the first, illustrated by the fowl and other game-birds, by the ostrich and by the ploceids, the castrated female, after a moult or artificial defeathering, puts on the striking male dress. A castrated male also puts on breeding dress, so that it looks as if this were the neutral form, which is inhibited by secretions from the ovary. This is confirmed by the injection of oestrogen into male fowls, which then acquire female plumage. In the ploceids pituitary hormones will induce castrates to put on male breeding dress at any time. The preparations used were mixed, but it seems likely that the active constituent was luteinizing hormone, which in mammals stimulates the production of progesterone.

The capon's plumage is slightly different from that of the cock, the feathers being brighter in colour and a little longer, and it is this, rather than strictly the plumage of the cock, that the castrated hen resembles. Apparently male gonadal hormones are slightly inhibitory, but less so than female hormones.

Ducks, both wild and tame, are similar; castrates retain the drake's nuptial plumage throughout the year, and do not, like normal males, lose it for a dull eclipse plumage for a few months in autumn.

The second type is found in ruffs and male black-headed gulls. Castrated birds do not acquire the breeding plumage, which therefore appears to be positively determined by the sex hormone of the testis. Thyroxine given to male house-sparrows tends to produce a female plumage, but in game-birds it induces male plumage.

5.32 Other epidermal structures

In some species there are different colours of beak in the two sexes, or one (usually the male) may have greatly developed appendages, such as the comb and spurs of the farmyard cock. These do not necessarily respond to hormones in the same way as the feathers.[32, 399]

Castration of the male fowl leads to a loss of the comb but the spurs remain. Castration of the hen also leads to regression of her much smaller comb, but she grows spurs. It seems therefore that the spurs are neutral and suppressed by female hormones, while the comb in both sexes needs the stimulation of a hormone to produce it. If a cock is castrated early enough it does not grow a comb, but will do so if it is then injected with testicular extract. The site of production of the hormone is the Leydig cells of the interstitial tissue. Oestrogen does not cause the growth of a comb in female castrates, and possibly it is normally induced by small amounts of male hormone produced by the persisting medulla, or male part, of the gonad.

The beaks of the starling and the black-headed gull change colour similarly in both sexes in spring; experiments on castration and injection of hormones show that in both sexes the change is mediated by androgens. That in the house-sparrow, where the female does not change, or changes only slightly, is also dependent on male hormone, as is that of the female Reeves' pheasant (*Syrmaticus reevesii*) where the male does not change. In the paradise whydah the female's beak changes from red to yellow, possibly under the influence of female sex hormones, and that of the male from red to black under the influence of the hypophysis.

5·33 The egg shell

The shell of birds contains much calcium, and they (or at least pigeons and fowls, on which experiments have been done) have made use of the property of oestrogen which affects its metabolism.[358] In various fish, frogs and snakes, oestrogen causes a high concentration of calcium in the blood. So it does in birds, but in addition it increases the absorption of both calcium and phosphate from the intestine, and causes the formation of bone in the marrow. These last two properties are shared by androgens. This accumulation of calcium phosphate in bone takes place during the early stages of the formation of the egg, for about 10–14 days, and in the later stages, when the shell is being formed, the bone in the marrow is destroyed, probably under the influence of the parathyroid, and its material used for the shell.

5·34 Behaviour

The importance of behaviour in the breeding cycle is discussed in the next chapter, but we may note here that behaviour and hormones are intimately connected, each having effects on the other. Male fowls or chaffinches that have been castrated cease crowing or singing, but can be stimulated to do so again by androgens. Conversely, song in budgerigars stimulates testicular activity.[47] Turtle doves (*Streptopelia risoria*) which have not themselves recently bred, when presented with young squabs, either attack them or court them; if they are injected with prolactin (which stimulates the crop glands) they react differently according to their past experience. Those that have bred feed the young, but those that have not ignore them, neither feeding nor approaching them. Shortly after the birds have laid eggs, but not before, the presentation of squabs to them induces crop growth without injection of prolactin.[151]

6

The Higher Life

6.1 COMPARISON OF BIRDS WITH MAMMALS

6.11 Instinct and intelligence

Birds show many complicated patterns of behaviour that are generally called instinctive. We will not attempt a definition of that much-abused word; all that it implies in its general use is that all members of a species (with due allowances for differences of age and sex) behave in a given situation in much the same way. In this use there is an obvious danger. All children brought up in the United Kingdom talk English, but if they were taken young to France they would talk French instead; so talking English is not instinctive, although a not very intelligent Martian might think that it was. In the same way, some of the things that all members of a species of bird do are learnt, not innate, and only careful experimental analysis can tell us the nature of any particular act. Some pieces of behaviour have components of both natures; when a bird migrates to Africa it is following an instinct, but when it returns not only to England but to the same nesting hole as it occupied the year before it is following an instinct and is also using knowledge of its surroundings that it has learnt.

When all this has been said, it remains true that birds show many patterns of innate behaviour, possibly more complicated and certainly more conspicuous than those shown by mammals. From this it has been customary to assume that birds are instinctive, mammals intelligent. An article published in 1964, for example, says that the extent to which the behaviour of birds can be modified or added to as a result of experience is very limited, whereas the behaviour of mammals depends to a larger extent on learning. Part of the object of this chapter is to show that this is nonsense.

Let us compare the blue and great tits with the house mouse (*Mus musculus*) and the wood mice (*Apodemus sylvaticus* and *A. flavicollis*). They are all of the same order of size, all are more or less omnivorous, and all are more or less commensal with man when the opportunity arises, feeding on what he provides, intentionally or unintentionally. The mice, especially the house mouse, enter houses regularly, the tits rarely. The mice attack almost any organic material, to which they are attracted presumably by smell. Tits almost certainly have little sense of smell and recognize things visually. Whereas man's food, such as a loaf of bread or a piece of turkey skin, almost certainly has much the same smell as grain or flesh, and so can be found instinctively by the mice, its shape is quite different from anything for which the tits could possibly have a built-in instinct. Each bird must therefore learn for itself, either by trial-and-error or by imitation of more experienced individuals, that these artificial or unusual foods are good to eat. That they do so, and do so very quickly, is obvious to anyone who keeps a bird-table.

Two special cases are well known. Monkey nuts (peanuts, ground nuts, *Arachis hypogaea*) were unobtainable in England during the war. In their shells they do not resemble any other form of nut, and by 1947, when they reappeared in the shops and were again put on bird-tables, no tit living in Britain could have seen one, yet the tits soon learnt that they were good to eat. In one series of careful observations[52] there was accidental discovery of their edibility by one great tit after the nuts had been displayed for five weeks, followed by imitation by its neighbours; (on this occasion blue tits were not seen to take the nuts). In the 1920s milk began to be left on doorsteps in bottles with cardboard stoppers. Both species of tit soon learnt to peck through the cardboard and eat the cream. The habit apparently started in two or a few centres, and spread throughout the country.[163] When the cardboard stoppers were replaced by aluminium caps the birds maintained the habit. Some simple experiments that I made in 1965 suggested that they react to several elements in the situation, for they would peck at caps of different colours, at full or empty bottles, and at a bottle stoppered with a champagne cork. No mouse has ever been known to learn in this way. If it be objected that while tits are probably amongst the more intelligent of birds, mice are amongst the stupidest of mammals, it may be answered that the orders to which they respectively belong, Passeriformes and Rodentia, are, judging by the number of species that they contain, the most successful in their class. And those who think that mammals are intelligent while birds are instinctive probably also believe that rats, which are scarcely separable generically from mice, are highly intelligent.

There are great specific differences in birds' ability to learn new ways of doing things. Tits and goldfinches (*Carduelis carduelis*) can learn to get food dangling at the end of a string by pulling up the string and holding it with their feet, but robins and chaffinches cannot. Garden blackbirds must often see their songthrush neighbours beating snails on the path to break the shells, but they seldom or never imitate them.

The study of animal behaviour, or ethology as it is now usually called, has acquired a vocabulary of its own that rivals that of psychoanalysis in its imprecision. The criterion of precision is that a good definition of a word can be substituted for the word itself without alteration of meaning. Judged by this test, the conceptual terms of modern ethology are unsatisfactory. For example, we are asked by one author to consider 'the question of the existence of a centrally directing drive'. But the same author later defines 'internal drive' as 'the complex of internal states and stimuli leading to a given behaviour', and there can be no doubt of the existence of these. So substituting the definition for the word does, in this instance, change the meaning. Other ethological terms, such as 'innate releasing mechanism' fare no better in such substitutions, so this sort of language will not be used in this book.

6.12 Innate behaviour

There is a clear theoretical distinction between behaviour that is innate and that which is learnt, corresponding in morphological terms to genotype and phenotype. 'Innate' must not be taken literally, for much behaviour covered by the term is not present at birth (or in birds, the equivalent of birth, hatching); it is acquired during ontogeny by a process of maturation, and is often if not always dependent on a preceding or parallel maturation of the appropriate organs. Sexual behaviour, for example, is not present in young birds, and chicks in the nest cannot fly; in the one case, gonads, and in the other, wings, are not yet fully developed. Isolated birds acquire most (but not all) of their sexual behaviour when they are old enough, and it was shown nearly 100 years ago that chicks of several species confined so that they could not flap their wings could fly at once when released at an appropriate age.[337] Flying and most sexual behaviour are therefore described as innate. The word 'instinct' is generally applied to behaviour of this type, and especially to its more complicated examples.

Whether a piece of behaviour is innate or learnt can only be found by experiment, which is seldom easy. Even when behaviour is learnt it still has an innate component; or, in other words, the genetic structure determines what can be learnt. Boys and girls can learn to speak French or English according to their upbringing, but apes, however well taught, can learn neither.

6.13 Learning[361]

Most if not all physiologists would agree that all learning depends on the formation of new lines of communication in the central nervous system, or on the suppression of those already existing, and, if this is so, at that level all learning is the same, but it is usual to classify it into types according to the sort of experience that is necessary for it.

One of the simplest ways of acquiring new habits is by habituation, learning not to do something that one would normally do. Some years ago the Ministry of Supply established in Wyre Forest a station for testing rocket-fuel. The ignition of the fuel makes one of the loudest noises imaginable which is unpleasantly audible miles away and continues for several seconds. At first the members of a tit-flock became immediately silent when they heard the noise, did not resume calling until a little while after it had ceased, and sometimes moved rapidly away. After a year or two they took no notice, and continued calling and moving steadily through the trees as if nothing had happened. They had learnt, as we would say in anthropomorphic language, that that particular sound, loud though it was, did them no harm.

Habituation is probably of some importance in the life of birds in simplifying instincts or in making them more useful. Spalding showed in 1873 that a newly hatched chick pecks at any small thing that is bright or moving, but it gradually learns not to peck at things such as its own faeces that are not good for food; its response is simplified to become more useful.[336]

It is sometimes difficult to distinguish in practice between habituation and what the physiologist calls fatigue, which also shows in behaviour as a waning of response. Fatigue should pass off, and then the response will return; unfortunately a response that has been lost by habituation may also return, and then we say that the animal has forgotten what it learnt. Perhaps at the neural level there is no difference between habituation and fatigue, but much fatigue is peripheral, in the sense organs or musculature, and is distinct from learning.

Habituation is essentially negative. The chief positive way of learning for most animals is probably by the formation of new, or conditioned, reflexes. A chick of a domestic fowl, as has been said above, has an instinct, or a reflex, to peck at anything small, bright and moving.[258] It has also a reflex to reject noxious substances, such as formic acid, placed in the mouth. If it is presented with a bee or a cinnabar caterpillar, it pecks at it, gets a new chemical stimulus or is perhaps stung, and spits it out. After a few trials (sometimes after only one) it will refuse the objectionable insect, and will also fail to peck at anything else that is similarly coloured, such as a drone fly. It has learnt that black and yellow things, such as bees, taste nasty; in other words it has acquired a new reflex, to reject things with that specific colour stimulus.

For the formation of a classical conditioned reflex (sometimes called Type I), as studied by Pavlov, certain conditions must usually be fulfilled.

The neutral stimulus (black and yellow in the above example) must be applied shortly before the effective one (formic acid in the sting); the neutral stimulus must not be one that already gives strong reactions. Occasionally both the conditions may be broken.

In another form of learning, sometimes called conditioned reflex Type II, new reflexes are formed as a result of a reward received after the action has taken place. This is the basis of trial-and-error learning. The opening of milk bottles and eating of peanuts by tits are learnt in this way. The birds have a natural reflex to peck at almost anything; in exploring, they discover that the milk bottles and the shells of the nuts contain food, which acts as a reward and so new specific reflexes are formed.

Finally new habits may be acquired without practice. There are two types of this. In latent learning the bird apparently responds to stimuli (or uses knowledge) that it has acquired but on which it has never previously acted, at least in quite the same way. A migrating bird that returns to its nest-site (or any bird that returns home after being displaced) is probably behaving like this. Experiments have shown that rats can learn to run a maze by random exploration without a reward, and finding a nest in a wood must be much the same as running a maze.

Occasionally, a bird may apparently see the way to solve a problem all at once, and perform the necessary action perfectly the first time; it is said to use insight. This is how goldfinches and tits appear to solve the problem of food on the end of a string.

The observations on which this classification of learning is based are derived from experiments on animals in simplified and controlled conditions. It is usually difficult to allocate a new piece of behaviour acquired in nature to a particular sort of learning with complete confidence. The rejection of black and yellow bees by chicks, classed above as conditioned reflex Type I, has also been regarded as Type II. Probably the learning is often mixed, and in any case the boundaries of the classes are ill-defined.

6.14 Instinctive or intelligent?

We can now return to the question of instinct or intelligence, and the comparison of birds with mammals. All birds and all mammals (including man) have certain pieces of behaviour that do not need to be learnt. If we like to call these instincts, then both birds and mammals are instinctive. If, on the other hand, we prefer to restrict the word instinct to more complicated pieces of behaviour, such as migration and nest-building, then birds show more of these than mammals, or at least show them more conspicuously. Similarly, we may use the word intelligence for all learning, or we may restrict it to those that are more difficult to explain, such as latent learning and insight, or even to the latter alone. In the wider view birds are certainly intelligent, for their responses are continually being modified. Even flight, though basically innate, is improved with practice; that is, its more difficult parts, such as landing, are to a great extent learnt, presumably by trial-and-error. On the more restricted view, that only insight learning indicates intelligence, then there is not much in birds. But neither is there much in mammals; it is doubtful whether any mammals except monkeys can equal the string-pulling exercises of some birds, and certainly only the higher Primates can exceed it.

If one looks in the other direction, and compares birds with reptiles, the differences are great. Although a few reptiles migrate and some show innate sexual behaviour above the level of copulation, they have none of the complicated instincts of birds. They can form conditioned reflexes, and possibly have some ability in trial-and-error learning, but the ease with which they achieve both of these is no greater than in the higher invertebrates. Claims that in knowledge of their territory they show latent learning are unconvincing, and they have no insight learning. Once again we see parallel achievements in birds and mammals, different in detail and reached by different methods, and with no similarity to other groups. It is largely because of this big difference in level of behaviour and learning that the birds are placed in a separate class from the reptiles, in spite of their structural similarity. In the rest of this chapter we shall consider some features of the higher life of birds, mostly instinctive but often with learnt components.

6.2 REPRODUCTIVE BEHAVIOUR

The complicated behaviour of birds shows itself best in connection with reproduction, and some of their striking habits, such as migration and territory-holding, that are not perhaps primarily reproductive, fit into the annual rhythm of which the central point is the production of young.

6.21 Pair formation

Birds in winter are in a neutral condition. They may be in flocks (which are discussed below) or they may be solitary, but they are seldom in pairs and sexual behaviour is absent. As spring approaches the flocks split up, or solitary birds come together, and pairs are formed. When this has happened the rest of the reproductive cycle follows—copulation, nest-building, egg-laying, and rearing of the young—until in the autumn the birds return to the neutral state. The exact pattern of pair-formation and the time at which it begins vary widely. Tits may begin it on sunny days in December, while late nesters, especially migrants, show no signs of it until May.

In most birds pair-formation takes place by a process which is conveniently called by the word courtship, using that word in the secondary but well-established English sense of 'wooing with a view to marriage'. Unfortunately ornithologists have tended to extend the word unnecessarily to include other forms of behaviour such as threat and pre-coitional foreplay, for which these alternative names are available. It will be used in the strict sense here, and regarded as one form of display which can be defined by its effect or (in teleological terms) its purpose. An alternative name for courtship is gamosematic display.

Courtship takes many forms, but the simplest is some form of showing-off by the male or the female or both. The cock blue tit raises his slight crest while facing the hen and shivering his wings. The goldfinch turns from side to side in front of the hen, flashing the gold of each wing at her alternately. It can hardly be doubted that we have here the function of the bright colours of many birds, and the most striking example of this sort of thing is the way in which the peacock (*Pavo cristatus*) having erected his tail, suddenly swings round to show its brilliant under-coverts to the beholder. As has been said above, colours are useless without colour-vision, and the only animals that have any behaviour comparable to this of birds (the teleosts, some apes, and a few insects) all have it.

The display of the tits begins while they are still in their winter flocks, and pair-formation is gradual. In a later stage the cock leads the hen to one possible nest-hole after another, pops inside, and then looks out, inviting her, as it were, to come in, which she may or may not do. When she does finally enter and accept a hole nest-building can begin and courtship is over.

Song may be important in courtship, especially in species such as the robin, chaffinch and most migrants in which the male selects and occupies a territory early in the process. The song, however, is a general invitation to any female who happens to hear it, and it is the female robin who courts the male by crouching submissively before him and refusing to be driven out when he attacks her, as he does all other members of the species who enter his territory.[194] The male whitethroat (*Sylvia communis*), having arrived and selected his territory some days before the female, accepts the first female that enters, flying at her and singing.[167] She responds by spreading her tail and flapping her wings, so that here courtship is mutual. There are chases in which either may follow the other. If, as occasionally happens, two hens arrive in the same territory, the decision between them is made not by the cock but by fighting between them until one is driven away. Male and female robins are alike, and the difference between the courtship of robin and whitethroat may perhaps be connected with this; only by her behaviour can the cock robin tell that the newcomer is a female and not a rival male. The whitethroats differ only slightly, the male having a greyer head, but this can be recognized by man and so presumably by the birds.

Other elements in courtship include the offering of nest-material and various forms of physical contact. The cock whitethroat may carry a piece of grass in front of the female. Some species caress each other with the beak ('billing and cooing' of pigeons) and in others small morsels of food are offered or given, usually by the male to the female. This 'courtship feeding' is much commoner as foreplay.

The most elaborate courtship is probably that of the bower-birds (Ptilonorhynchidae) of Australia and New Guinea.[233, 235] They are passerine birds living in forests and making nests in trees. The male, strikingly coloured and different from the female, builds a platform, bower or avenue, according to the species, of twigs, moss and so on, and decorates it with conspicuous objects such as flowers, beetles' elytra and bleached bones. He attracts a female to the bower, and displays to her, picking up some of his collected jewels and calling with much mimicry. Some species paint the walls of the bower with charcoal or fruit pulp moistened with saliva, using a wad of fibre as a brush. It is some months before the female copulates and builds a nest, sometimes hundreds of yards from the bower.

A number of birds, notably the ruff and reeve and some of the game-birds, have a type of courtship called the lek, a name which strictly belongs to the blackcock and greyhen (*Lyrurus tetrix*). In these species the males in spring show a striking difference in colour from the females (hence the separate English names for the examples given), and assemble in well-marked places for a few hours a day to dance and display, each cock holding a small area of ground. The females cluster round, and apparently exercise a real choice in the males with which they will pair. Polygamy is normal, some cocks getting several mates, some few, and some none. In the American sage grouse (*Centrocercus urophasianus*) there may be as many as 400 cocks at a lek, but 87 per cent of the matings are achieved by fewer than three per cent of them.[42]

At some ruff leks, especially small ones, there are 'satellite' cocks, which differ in colour from most, do not display or hold space, but are tolerated by those that do. They nevertheless achieve copulation.

In species of the Brazilian manakin of the genus *Chiroxiphia* a group of males display communally to a female, but pairing does not follow. For it, there is a silent display of one cock to one hen.[42]

The time for which a pair of birds stay together varies. Usually it is one season, but numerous examples of pairing for life are known, such as the blue tit and house-sparrow.[326] It is sometimes impossible to know whether there is really a firm marriage that would withstand assaults from other individuals, and how far the formation of the same pair that occupied a territory the year before is due to an attachment to the old nest-site in both sexes. Where the species migrates and the chance of the separation of the pair is high the second would seem likely. At the other extreme, lek birds do not usually form pairs at all, since the display leads merely to copulation, after which the sexes go their own way, the male continuing to visit the lek and attract other females, while the hen departs to make a nest and lay and incubate her eggs. The male is promiscuous rather than polygamous, and in this resembles many lower vertebrates and many mammals such as rabbits and grey seals (*Halichoerus grypus*).

The display of courtship is often similar to that of precoitional foreplay, or epigamic display, and in the lek birds the posturing serves both purposes. Sexual display and play continue more or less throughout the breeding cycle, being usually most marked before the eggs are laid. Feeding of the female by the male often continues when she is sitting, for example in the red-backed shrike (*Lanius collurio*) and hawfinch (*Coccothraustes coccothraustes*), and sometimes even after the eggs have hatched. In this case, and in the two species just mentioned, the cock may give the food which he has collected for the chicks to his mate, and she then passes it on. This might seem a possible origin of the symbolic presentation of smaller quantities of food to the hen during incubation or before coition, but this is only guesswork. Certainly the behaviour of the female receiving the food often resembles that of the chicks, and the robin and others give calls similar to those of the young. The male too feeds her in a similar way to that which he uses for the chicks, pigeons for example regurgitating their milk. The insect-eating pied flycatcher (*Muscicapa hypoleuca*) leaves the nest more often if she is not fed than if she is, so that the feeding here is functional in a straightforward way, and not symbolic.[145]

Perhaps more surprisingly, courtship somewhat resembles aposematic or threatening display. Elaborate attempts have been made to dissect display into components according to the gestures adopted and to relate these to what are called tendencies—to flee, to fight and so on. These explain nothing, and merely restate the phenomena in other, and less precise, words. We know that sexual behaviour and aggression both depend on hormones, especially adrenaline and those of the gonads. It is possible that the choice of one sort of display or another depends in part on the balance of these hormones, but the effect of a hormone itself depends on the state of its target organ; that in turn will depend, amongst other things, on its previous and especially its recent history, including the hormones and external factors that have acted on it. Some time it may be possible to give a causal analysis of display in some such terms as these, but at present we must leave the facts and their effects to speak for themselves.

6.22 Nests

Not all birds make nests, if by that word we mean some specially built structure, but all except the emperor and king penguins (*Aptenodytes forsteri* and *A. patagonica*), which carry their eggs around on their feet, lay their eggs in some definite spot. The selection and maintenance of this are usually included in the term nesting, even when no nest is built.

The typical nest is a cup of twigs, grass or other vegetable material, either on the ground or suspended in a bush or tree. In non-passerines it is seldom more than this, and in some even less, for ostriches and nightjars (nighthawks, Caprimulgidae, Caprimulgiformes) lay in a mere scrape on the ground without any added material, and auks (Alcidae, Alcae) lay on bare cliff ledges. Most passerines line their nests using a different and softer material, such as moss or feathers, or mud. The ducks and geese (Anseres) are exceptions to the rule that non-passerines make simple nests, and are the only birds that use feathers plucked from their own breasts for lining. Several passerines, such as some wrens (Troglodytidae), leaf-warblers (*Phylloscopus*) and the long-tailed tit (*Aegithalos caudatus*), build domed nests, with the entrance at the side. The extreme of complexity in nests made of vegetable materials is in the old-world weavers (Ploceidae) and new-world Icteridae, which, as the English name of the former implies, can thread filaments of grasses into a sort of cloth. Some species can even make half-hitches.

A few species use plastic material. Many swifts (Apodidae) use the mucin of their own saliva, with a greater or lesser amount of vegetable material or feathers to make a cup which may be attached to the side of a a rock or cave, and in some species of the genus *Collocalia* there is little else but saliva and soup can be made from the nests. Some members of the Hirundinidae, such as the European house-martin (*Delichon urbica*) and the American cliff swallow (*Petrochelidon pyrrhonota*), make nests of mud; both these species stick them to vertical surfaces of cliffs, and both have taken to nesting under the eaves of houses.

Others use holes, either in the ground (wheatears; petrels, Procellariiformes), in a sandy bank (sand-martins, *Riparia riparia*; kingfishers, Alcedidae; bee-eaters, Meropidae) or in trees (woodpeckers, Picidae; many tits). They may excavate these for themselves, or make use of a pre-existing hole. Of the species-pair marsh tit and willow tit (*Parus palustris* and *P. montanus*), which are hardly distinguishable by sight in the field, the former scarcely ever excavates its own hole, while the latter always does. None of the non-passerine hole-nesters mentioned lines its nest, but passerines always do.

6.221 *Choice of site*

The first step in nesting is the search for the site. This may begin, as in some tits, in the previous winter or even autumn, although since these birds roost in holes similar to those in which they nest some of the observations made in September may be merely of birds looking for somewhere to sleep. By December at least there can be no doubt that the birds are looking for a nesting site, the male leading in the search and the female making the choice. Sometimes lining material is placed in several holes, but finally all but one are abandoned. Birds that choose their nest-sites as early as this will not use them for some months. At the other extreme late migrants may choose a site and start building within 24 hours. In general, in British birds, the sexes co-operate in searching for a site, but the female has the final say in its situation.

It is obvious that a bird that nests in a special sort of place such as a hole is more restricted in its choice than one that nests amongst the twigs of a bush or on the ground, but even birds that use places such as these may show an attachment to a good nest-site once it has been found. This is best known in the golden eagle (*Aquila chrysaetos*), some of whose Scottish eyries have been occupied for centuries. It is generally assumed that in this sort of case either the members of a pair remain mated in successive years, or one member of the pair has died and the other has found a new mate, so that either way there is memory and continuity, but the case is not as simple as this. An eyrie may be unoccupied for a year or two, after which the birds return; it looks as if some rocky ledges are specially attractive to the birds. The same sort of thing can happen with other species. Blackbirds sometimes build two or three nests on top of one another in successive seasons. A few years ago I was sitting with an ornithologist in the garden of some friends of his parents whom he had not visited for 30 years. He suddenly said 'when I was a boy there always used to be a goldcrest's nest in that cedar', got up, crossed the lawn, put his hand under the bough and flushed a goldcrest (*Regulus regulus*) off its nest. This bird usually attaches its nest below the branch of a conifer, but sometimes uses other sites, such as ivy. In the district where this incident occurred the bird is common, and its natural nesting-place is in the isolated yews in the near-by oak forest; but there are extensive Forestry Commission plantations a mile away, which should provide plenty of sites.

There is some slight American evidence that, even when it uses different sites, an individual persistently chooses one that is about the same height from the ground.[41] This is a refinement of the fact that each species of bird has a limited range of types of site, but that within that range some individuals or populations have persistent choices. The peregrine falcon (*Falco peregrinus*) in Britain nests almost solely on cliffs, on the continent often in old nests of other species in trees, and in America frequently on buildings. The jackdaw nests in a variety of situations, but chiefly on cliffs in England and in trees in Scotland. Many species make occasional use of man's buildings and walls, and a few, such as the house-martin, swallow and swift (*Apus apus*), are now almost confined to them.

Attachment to the nest-site begins, even in migratory species, with return to the home area. This happens, though not very markedly, with year-old birds, some of which can be found nesting near where they were hatched and bred. That more of them do not appear may be due to the occupation of most of the suitable places by older birds, which in many species show a strong tendency to return near where they nested in the previous year. In the pied flycatcher about a third of the adults ringed at nest-boxes in one year are found in the same area, though often in different nest-boxes, in the next year.[147] Since the annual mortality is about 50 per cent, this means that about two thirds of the adults return (Fig. 6.1 and 6.2). There is some geographical variation and in Finland, though the proportion of males that return is the same as in England and Germany, that of females is much less. In the Forest of Dean the males return to a nest-box at an average of 125 yards from that last occupied, the females to one 175 yards away.[55] Faithfulness to the home area is clearly stronger in males than females. This is perhaps to be expected since the male chooses the nest-site before the female arrives. Pied flycatchers that have once returned tend to do so again, so that faithfulness appears to be an individual characteristic. The same is true of the American song-sparrow.

6.222 Nest building

After the site is chosen, the nest, if there is one, may be built. Usually the female does this alone, but in many species the male helps, and in a few he may do it all. Twigs, grass, moss, or whatever may be appropriate, are collected and placed in position, and as the cup begins to take shape the bird sits in it and presses it into a compact structure. Only in the weavers and a few others is there any formal arrangement; in most the result is what would be achieved by random placing of material to make a tangle. In plastic nests there is a more consistent placing of the material, first to stick the foundation to the cliff or whatever it may be, and then to build up the walls.

Besides the main structure the nest may have an outside decoration, as in the chaffinch and long-tailed tit, both of which use lichen. When the chaffinch uses this on a nest on or against a branch of a tree also covered with lichen, as is often the case in northern or western woods, the nest is completely indistinguishable from the bough when seen from the side or below, and one may guess that this is the natural habitat of the bird in which the decoration has been selected. In other situations, as in smooth-barked trees, the lichen makes the nest more conspicuous, and this is even more so when, as sometimes happens, odd material such as paper is used for the decoration.

When nests are lined there may be a change-over in building to the use of some special material such as mud in the song-thrush and feathers very commonly. The long-tailed tit uses about 2000 of these, collected one

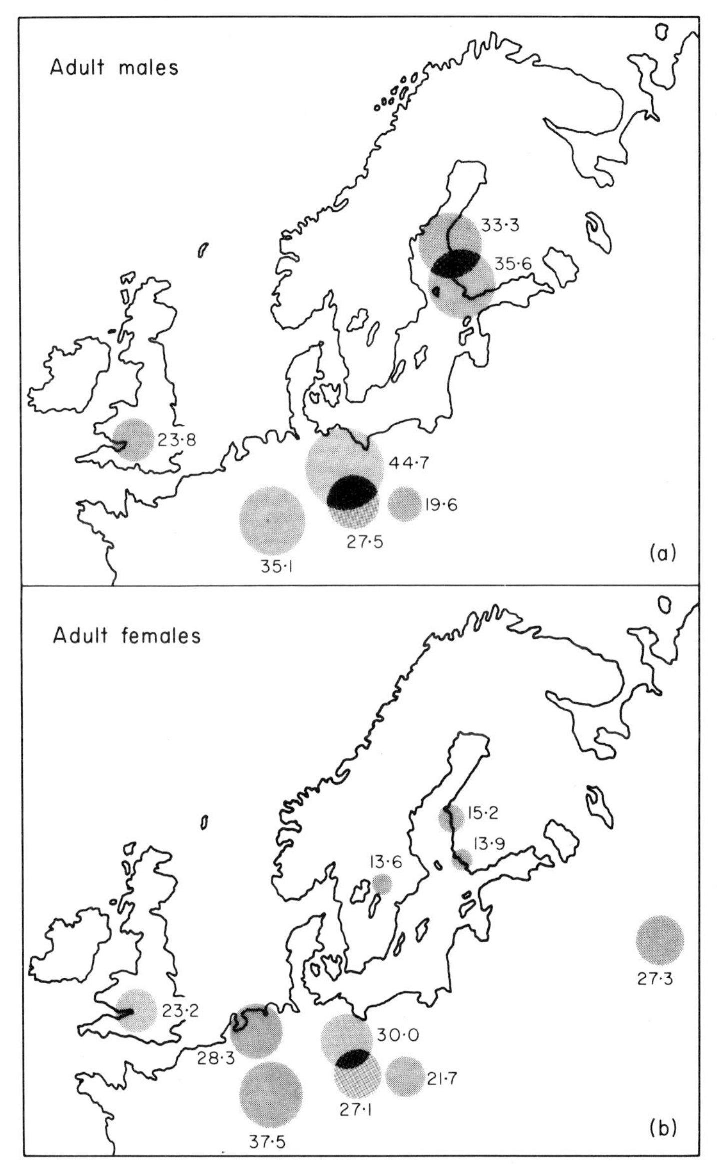

Fig. 6.1 Return of the pied flycatcher (**a**) male, and (**b**) female, in different study areas. The diameters of the circles are proportional to the return ratios, shown as numerals on the diagrams. (From L. von Haartman, 1960, Figures 1 and 2, *Proc. XII int. orn. Congr.* **1**, 268–9.)

by one, which must represent a considerable and wasteful expenditure of energy, since other birds can get on without them. The only considerable natural source of such large quantities of feathers would seem to be the bodies of birds killed by sparrow-hawks and other predators, which may in part explain the relative rarity of the long-tailed tit.

It has been known since at least the mid-nineteenth century that domesticated canaries build their nests by instinct, for fanciers regularly removed the nest and substituted a felt cup; yet birds hatched and reared in these cups built proper nests when they grew up.[303] The building process is similar in some ways to the instinctive construction of a web by a spider. Here things have to proceed in a certain order, the doing of one thing, such as the making of the radii, acting as a stimulus for the doing of the next, such as the making of the spiral. In the same way the lining of the nest follows automatically after the building of the cup, either because, as in the spider, the doing of one action is itself the stimulus for the next, or because the sight of the half-built nest initiates the completion. It has been claimed that in domesticated canaries the latter is the explanation, but that the change-over from the use of grass to the use of feathers depends also in part on a change in the internal state of the bird.[162] Experiments on the American tricoloured redwing (*Agelaius tricolor*) showed that if the lining of the nest was removed there was no repair; if building was interrupted by natural causes the bird laid in an uncompleted nest.[98] More often, if a nest is disturbed, the bird gives up and starts again elsewhere.

6.223 Nest sanitation[36]

One of the difficulties of young being hatched and kept in the nest for some time is that faeces may accumulate. In most non-passerines the only habit that has been acquired that deals with this is that as soon as they can move the chicks go to the edge of the nest and defaecate over the rim. Similarly brooding gannets (*Sula bassana*) face inwards on their nesting ledges, perhaps because this enables them to defaecate into the sea. The nests of many hole-nesting non-passerines, such as kingfishers, become very foul.

Older nestlings of passerines may also defaecate over the edge of the nest, but most species have two other special features: the faeces are enclosed in a membranous capsule, so that with care it can be picked up; and for the first few days after hatching these capsules are swallowed by the parent, especially the female, often being taken direct from the vent of the nestling. Later they are carried away and dropped; this is very easily observed for example when starlings are nesting under a roof and feeding young. At almost every journey they carry away a capsule and drop it in flight about 20 yards away. Woodpeckers (Picidae) also swallow or carry away the faeces.

6.23 Brooding

After nest-building, the next type of instinctive behaviour is brooding the eggs. The function of this is to keep them at the correct temperature, for they will not hatch if they are much colder or much hotter, and perhaps to a lesser extent to provide suitable humidity. In this birds differ from all reptiles, and indeed from other animals.

The onset of brooding, well known as a special state in the domestic hen, involves a physiological change in the bird, which is marked in the majority of species by the development of brood patches, areas of the abdomen which lose their feathers and become highly vascular.[22] In passerines they coincide with the ventral apteria, so the only feathers to be shed are the down. The size and number of the blood vessels increase, and there is rapid cell-division in the epidermis. These changes begin before the first egg is laid and last throughout incubation. The skin has returned to normal by the time the young are ready to fly, but the down is not restored until the autumn moult. The brood patches are pressed against the eggs, and presumably heat is thus more readily conveyed to them. In the few species that have been investigated, including the canary and some American finches, brood patches can be induced in non-breeding birds by injection of oestradiol (an oestrogen) and prolactin, or, if the pituitary is intact, of the former alone. Their development is presumably in general part of the endocrine cycle. In some birds, such as cormorants and gannets, brood patches do not occur, but this sort of minor deviation from the usual sexual cycle can be paralleled in other ways, and in mammals. Usually only the female broods but there are many birds, such as some species of swallow and especially non-passerines, in which both species do so, and in these the male usually has brood patches.[322] There is, however, no exact correspondence between brooding and the possession of patches. For example, some species of the American genus of flycatchers *Empidonax* have male brood patches, some do not, but in none does the male incubate. Males of the bush-tits (*Psaltriperus*) do incubate, but have no patches. It looks as if brooding by both sexes was primitive, but has been lost many times, sometimes long ago, sometimes recently, as in *Empidonax*, and has been reacquired in a few birds such as *Psaltriperus*. As well as the appropriate internal state the bird may need the stimulus of sight or touch of the eggs to start brooding, and if this is not provided the impulse may pass off, or if the bird is presented with chicks from another individual it may feed these and miss out the incubation period. Incubation may sometimes be stimulated more strongly by an artificial stimulus than by the normal one.[367] Thus an oyster-catcher (*Haematopus ostralagus*) prefers an abnormal clutch of five eggs to the normal three, and an egg four times the normal length to one of its own. A ringed plover (*Charadrius hiaticula*) prefers a white egg with large black dots to its own spotted brown ones.

Brooding may include other types of behaviour than merely sitting on the eggs. When the bird leaves the nest to feed it may cover the eggs with vegetable matter, as do grebes (Podicipedidae) and moorhens, while some ground-nesting species such as the black-headed gull can retrieve an egg that has moved a short distance from the nest by rolling it back with the beak.

6.24 Non-nesters

A few birds neither make nests nor incubate. Some game-birds of the family Megapodiidae[117] bury their eggs and leave them to be hatched by the heat of the sun, so having reverted to the custom of many reptiles. Others in the same family make mounds of vegetation which provides heat by its fermentation, and in these mounds the eggs are buried. The male attends the mound, probes and presumably tests it, and by stirring the vegetation and exposing the eggs, which may cool or warm them according to how far they are exposed to the sun, maintains them at an approximately constant temperature.

Many cuckoos (Cuculidae), cowbirds (Icteridae), and honey-guides (Indicatoridae), some weavers (Ploceidae) and a few other birds, are brood-parasites, laying their eggs in other birds' nests. The origin of this habit is difficult to explain. One cowbird (*Molothrus badius*) is not parasitic; it sometimes builds its own nest, but more often uses the nest of another bird, sometimes an abandoned one and sometimes an occupied one from which the owners are driven out. Several other species of bird will evict others from their nests—house-sparrows will drive out house-martins, and starlings will drive out great spotted woodpeckers (*Dendrocopus major*). It is perhaps a small step from driving out the owners to laying eggs in the nest without doing so. The skin of cowbirds does not form brood patches when prolactin is injected.[165]

Whatever its origin, brood parasitism involves some of the most complicated and strange instincts in the animal kingdom. European cuckoos (*Cuculus canorus*)[58] and North American brown-headed cowbirds (*Molothrus ater*) watch their host's nest being built, and wait until the appropriate moment, when one egg or more has been laid, to deposit their own. Some species parasitize a wide range of host (though, as in European cuckoos, an individual may be much more restricted) but others are limited to one or a few hosts only. Each species of whydah and combassou (Estrildidae, Ploceidae) has a single host belonging to the same family, and the males imitate the song of this host, as well as having their own. The females only ovulate when they see an appropriate host building its nest.[274]

The nestlings of European cuckoos and some others have evolved the habit of ejecting their foster-brothers from the nest. This presumably has survival-value (provided the number of parasites is small compared with that of the host), but the habit by which the hosts preferentially feed the cuckoo and neglect their own young is difficult to explain.

7

Maintenance Activities of Reproduction

7.1 SONG[18, 154, 362]

The application of the word 'song' to some of the sounds made by birds is an example of the transference of a word for a human artefact to a natural production; its use in this sense occurs in Old English and in other languages and is at least as old as the Song of Solomon (3rd century B.C.). Musical sounds, to which the word is strictly appropriate, are found only in the suborder Oscines of the order Passeriformes and in a few other birds such as some pigeons, but ornithologists have extended its use to other vocal sounds that appear to have a similar function, and even to mechanical sounds, such as the drumming of woodpeckers, that serve the same purpose. These latter are better called song-substitutes.

No strict definition of bird-song is possible, but a working definition might be that it is any sound or series of sounds of a recognizable pattern produced solely or mainly in connection with pair formation, actual or potential. Generally it is associated with the holding of territory (see below) and where it occurs outside the breeding season, as in the robin, this is nearly always so. Generally only males sing, but females may do so where they hold winter territory (again the robin is an example), and in several tropical birds both sexes sing. Good song is associated with relatively dull colouring and with little or no sexual dimorphism.

7.11 The structure of song

The sounds of birds are produced by the syrinx, and the quality of the songs of Oscines may be correlated with the complexity of this organ mentioned in Chapter 3. Songs may be classified in musical terms, and cover a range as follows, though there are intermediates and no hard lines can be drawn.

1 A single note or a few notes of low musical quality; e.g. the brambling (*Fringilla montifringilla*) and many tropical species.
2 A simple phrase of one or a few notes of higher musical quality; e.g. the cuckoo, chiffchaff (*Phylloscopus collybita*) and many doves. There is usually occasional variation such as the stuttering of the cuckoo to 'Cuck-cuckoo', and the change of the chiffchaff's normal two-note phrase to a four-note 'chivvy-chaffy'. No non-passerines go beyond this.
3 A longer song of the order of 10 notes, in which there may be noticeable variations between individuals and less marked variation between successive songs of the same individual; e.g. the chaffinch, the song sparrow, and many buntings (Emberizidae).

   Songs of these three types, in which there is little variation, are usually repeated many times at short intervals.
4 A number of phrases which may be repeated and put together in a very large number of ways, so that hardly any burst of song of 10 seconds or more is an exact repetition of any other; e.g. most thrushes (Turdidae). The nightingale (*Luscinia megarhynchos*) comes here, but differs from most of the others in that there is a much wider range of quality between the different phrases.
5 A continuous song in which the human ear can recognize few if any repetitions; e.g. the skylark (*Alauda arvensis*), and, less certainly, some warblers such as the blackcap (*Sylvia atricapilla*) and garden-warbler (*S. borin*). It is probable that this type is really assimilable to 4, but that its speed and absence of pauses makes the phrases of which it is built up indistinguishable.

The notes of which bird-song is composed, like the notes of a musical instrument, may be fully expressed by four qualities or parameters. First, the pitch, which is determined by the lowest frequency (called the fundamental) of the vibrations that cause it; the higher the frequency the higher the pitch, and doubling the frequency raises the pitch by one octave. Second, the loudness, which is determined by the amplitude of vibration of the source, which in turn determines the rate at which energy is supplied to the human ear drum (i.e. the power). The ear has a built-in volume-control, so that perceived loudness bears no simple relation to amplitude. Third, the timbre (by analogy sometimes called tone-colour), which is determined by the frequencies higher than the fundamental (harmonics, overtones, upper partials) that are simultaneously present. Very few sounds are pure, that is consist of the fundamental only, the nearest being those produced by a tuning fork; good boys' voices are nearly pure. In musical sounds the frequencies of the overtones have simple arithmetical relationships to that of the fundamental, while in noises there is no such relationship, and some of the overtones may be louder than the fundamental. Lastly, the duration of the note is important.

Man recognizes a piece of music by all four characteristics, but the most important, at least for the non-musician, are the duration of the successive notes and the relative (but not the absolute) pitch. We would recognize a tune as the National Anthem whether it were sung by bass or treble, or played by a recorder or an orchestra, but if the timing were altered we might not recognize it. There is some evidence that birds are similar.

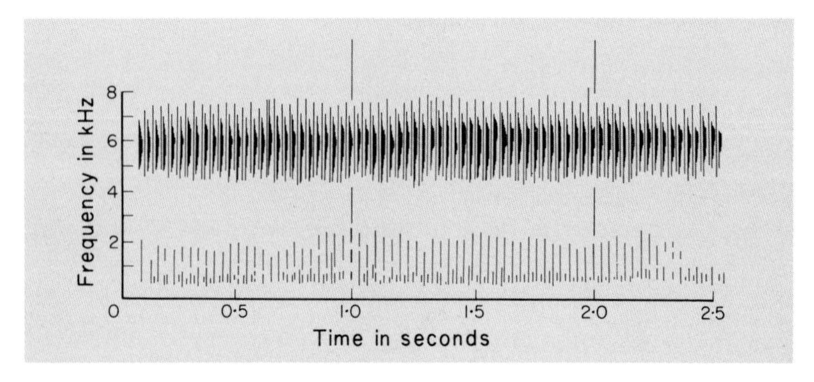

Fig. 7.1 Song of the grasshopper warbler; it has 31 triple pulses per second with an energy peak at 5 kHz. (From Thorpe,[362] 1961, *Bird Song*, Fig. 51(a), University Press, Cambridge.)

The poor songs of birds are noises, with random and loud overtones. The good songs have more regular harmonics, but still usually more than most musical instruments. Moreover, in going from one note to the next, birds often sing *portamento*, that is they slide up or down the scale, forming all the intervening notes as they go. This is rare in music, though it can be achieved by the human voice, the trombone, and bowed stringed instruments such as the violin.

Most of the qualities of bird song can be detected by an attentive human ear, but study has been greatly helped by the invention of the sound spectrograph, an instrument which analyses sounds for all four qualities and produces a record on paper. Pitch (including overtones) is shown by the position of the trace on a vertical scale, duration by the horizontal scale, and loudness (rather less clearly) by the darkness of the trace. Examples of sound spectrograms are shown in Figs. 6.3–6.8 and 6.10–6.13.

The trill of the grasshopper warbler (*Locustella naevia*, Fig. 7.1) shows a rapid succession of notes, about 30 per second, with two main components, both with a range of a few hundred Hertz. Each of the upper and louder notes is double, with a short grace-note in front of the main sound. The song has low musical quality.

The corn bunting (*Emberiza calandra*, Fig. 7.2) has a song which is slightly more musical, but still with a wide range of frequencies in each note.

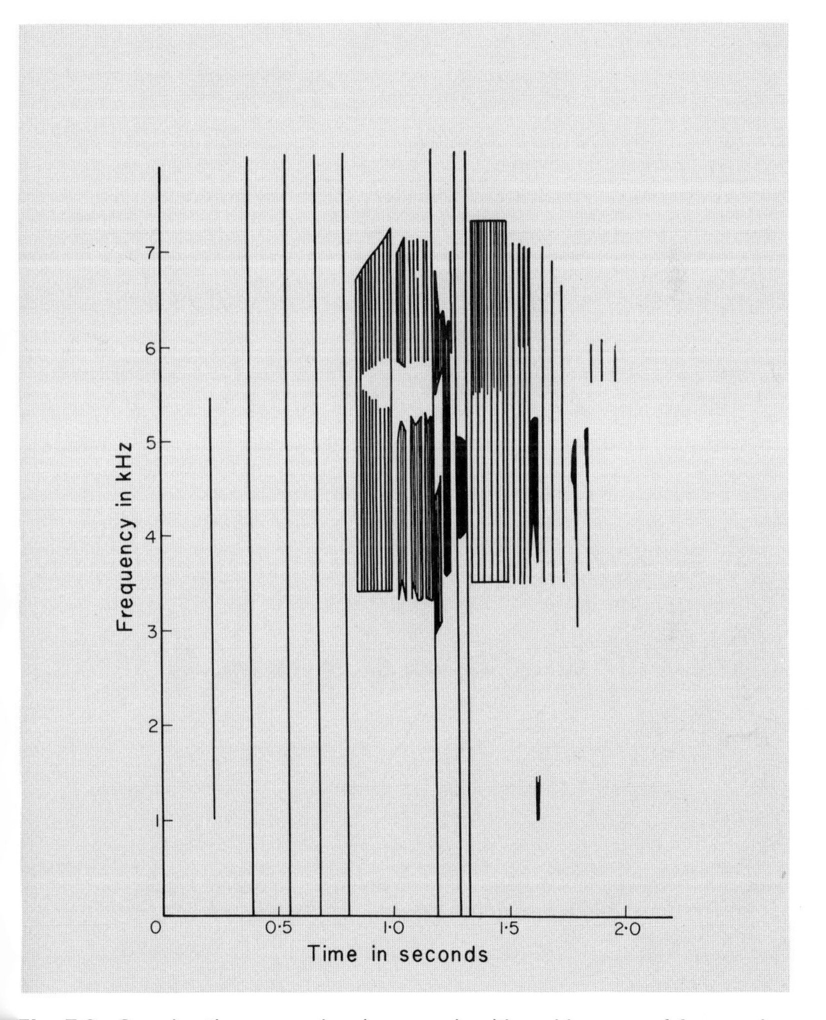

Fig. 7.2 Corn bunting song, showing sounds with a wide range of frequencies and very short duration, often heard as clicks. (From Thorpe,[365] 1961, *Bird Song*, Fig. 26, University Press, Cambridge.)

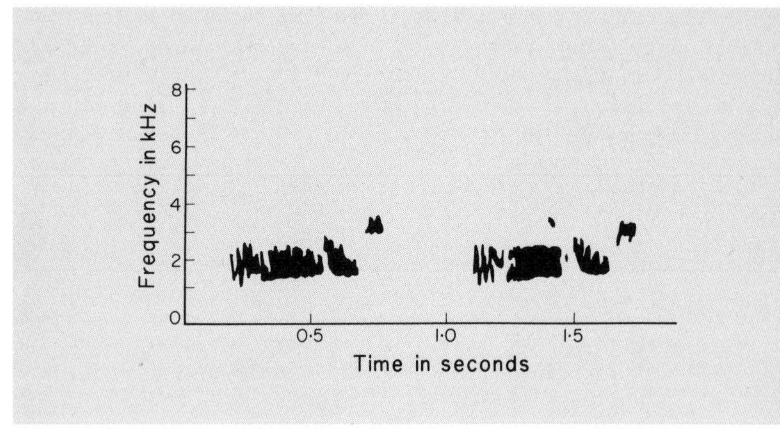

Fig. 7.3 Typical phrases from the full song of a blackbird, *Turdus merula*, N London, April 1957. The phrases can be approximately rendered by the syllables *tll-ew, tll-ui.* Note fairly pure tone and the fundamental frequencies of the characteristic notes restricted within a range of about 1·5 kHz with little sound above 2·5 kHz. (From Thorpe and Pilcher,[365] 1958, Fig. 1, *Br. Birds* **51**, 509.)

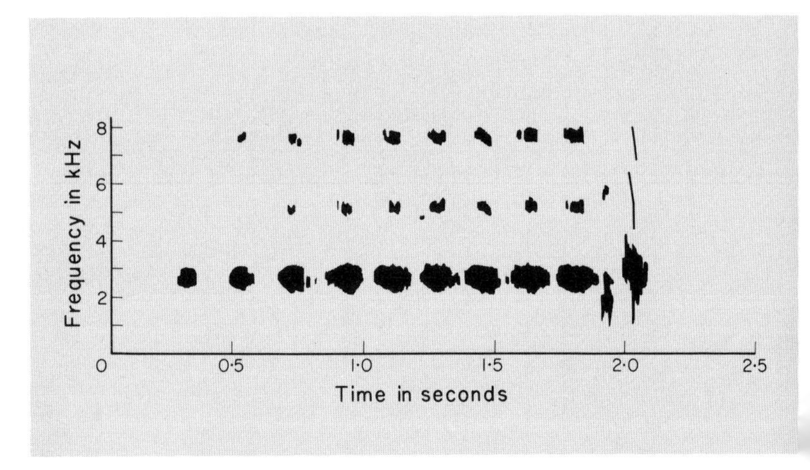

Fig. 7.4 Note from the song of a nightingale, showing regular harmonics. (From Thorpe,[362] 1961, *Bird Song*, Fig. 28, University Press, Cambridge.)

The blackbird (Fig. 7.3) and the nightingale (Fig. 7.4) have notes of a much higher musical quality. In the two phrases shown for the former the spread of frequencies is comparatively low, and the second note of each shows the characteristic *portamento*. The nightingale's crescendo on one note (pure Sibelius) shows a similar narrow range of the fundamental, but there are also other notes at twice and three times its pitch, i.e. at the octave and at the fifth above this; these are the second and third harmonics (the fundamental being the first) of the musician.

Figure 7.5 of the brambling illustrates the limitations of the sound

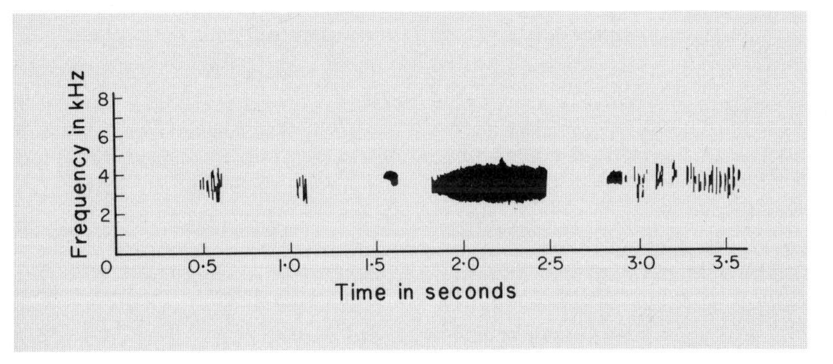

Fig. 7.5 Song of a brambling, recorded in Sweden. (From Thorpe,[362] 1961, *Bird Song*, Fig. 56, University Press, Cambridge.)

spectrograph. The first three notes are very faint, and are seldom heard, but are followed by a loud one that in the spectrogram does not differ much from the first note of the blackbird in Figure 7.3 except for being a fifth higher. It is, however, quite different when heard, and is one of the harshest bird sounds in the northern hemisphere.

The robin (Fig. 7.6) is an example of a complex song, several separate

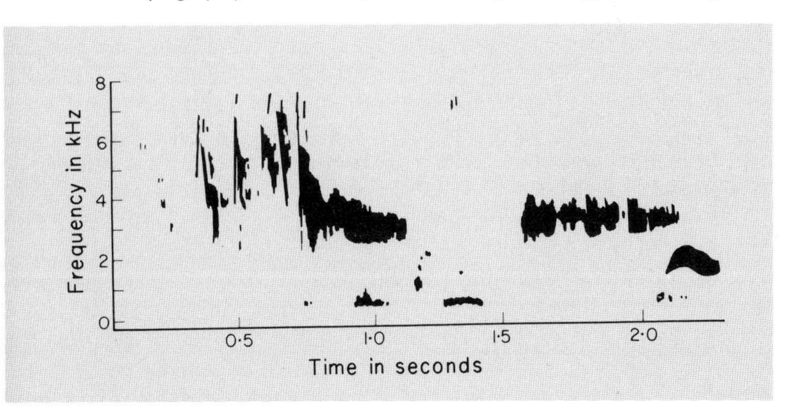

Fig. 7.6 Robin's song, showing high front sounds and extensive slurring. (From Thorpe,[362] 1961, *Bird Song*, Fig. 27, University Press, Cambridge.)

notes covering a wide range of frequencies, and with much *portamento*.

7.12 The function of song

Typical song is given by the male in his territory, and is sexual in character, brought on as part of the seasonal and endocrine breeding cycle described in Chapter 5. Many observers have recorded how song, loud and frequent when the territory is first occupied, declines in both respects when a female has arrived and the singer is mated. This is well shown in Von Haartman's study of the pied flycatcher in Finland (Fig. 7.7). The decline in the later part of the breeding cycle, when young are being fed, is obvious to the most casual observer of garden birds, and by July only a few late breeders, such as the yellowhammer (*Emberiza citrinella*), are singing in England.

Many birds sing also in other sexual situations; for example, wrens sing in courtship or foreplay, and the brown thrasher (*Toxostoma rufum*) and many finches during selection of the nest-site. A few species, such as some swallows, even sing on or near the nest, while incubating or feeding the young. In all these cases the song is often recognizably different from the normal territorial song, and is usually much less loud.

When birds sing outside the breeding season they are often holding territory; for example, both male and female robins sing and are territorial in winter. In other species singing may be associated with partial sexual stimulation. On bright days in winter, for example, skylarks may rise from the flock and sing briefly. In many migratory species there is a slight increase in sexual hormones in connection with the autumn departure, and some, such as the willow-warbler, sing a little on their way south.

Most singing hens are probably out of the normal hormone balance, since they usually behave like males in other ways, for example in holding territory. Females of some species, such as canaries and chaffinches, have been induced to sing by injection of testosterone or its derivatives, and aged blackbirds and domestic hens have been known to sing with partial sex-reversal.

In a few species, the two members of a pair sing a duet, either in unison or, more often, with different phrases sung antiphonally. Such duets have been chiefly described in tropical birds, but the tawny owl is a good British example. The female, unusually, calls first, singing 'Kewick', and is immediately followed by the male's 'too-hoo'. This song is distinct from the male's normal territorial song, which is longer and has two phrases; the first consists of two hooted notes, a rest of five seconds follows, and then comes the second phrase of five hooted notes of which the first is very faint. The male's phrase in the duet is somewhat similar to the first

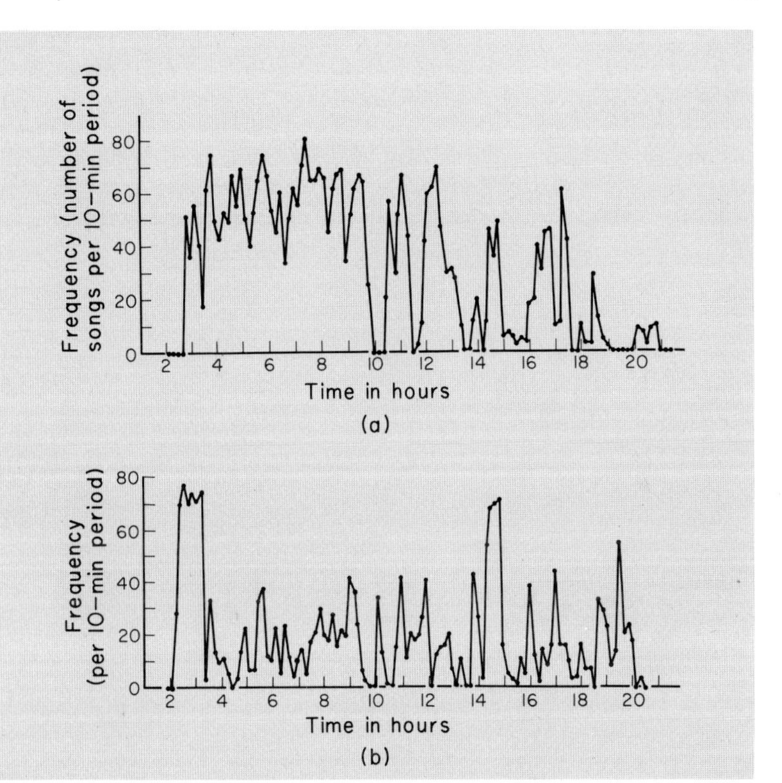

Fig. 7.7 The song rhythm of the pied flycatcher. (**a**) The curve shows the total song performance of an unmated male during the 24 hours of 3rd June 1955, Tvärminne, Finland. The curve is typical of unmated males. (**b**) The curve shows the song performance of a mated male during the 24 hours of 1st June 1959, Tvärminne, Finland, a day in which his female was engaged in nest-building. The curve is typical of this situation. Note the high frequency in the early morning before the female is awake. As soon as she appears the male is much less vocal. (From Thorpe,[362] 1961, *Bird Song,* Fig. 23, University Press, Cambridge, after L. von Haartman.)

phrase of the territorial song. This duet is presumably the origin of the Shakespearean 'tu-whit, tu-who'.

A few forms of singing, such as that of autumn skylarks or willow-warblers, or of swallows near the nest, seem to have no utilitarian function, and appear to be examples of some form of emotional expression, but in the commoner types song is clearly a form of communication. Its specific distinctness is therefore to be expected. For the most part, the song of any species of bird can be easily recognized by a man with a moderately good ear after a little practice, and general observation, as well as the play-back of tapes, shows that birds respond to the song of their own species. Even where there is little exact repetition, as in the thrushes, the general pattern, pitch and timbre of the song are specific, and may be recognized rather as one may switch on the radio and say 'Mozart' without being able to decide which particular work of his is being broadcast. There are, however, some puzzling overlaps which have not been adequately explained or even

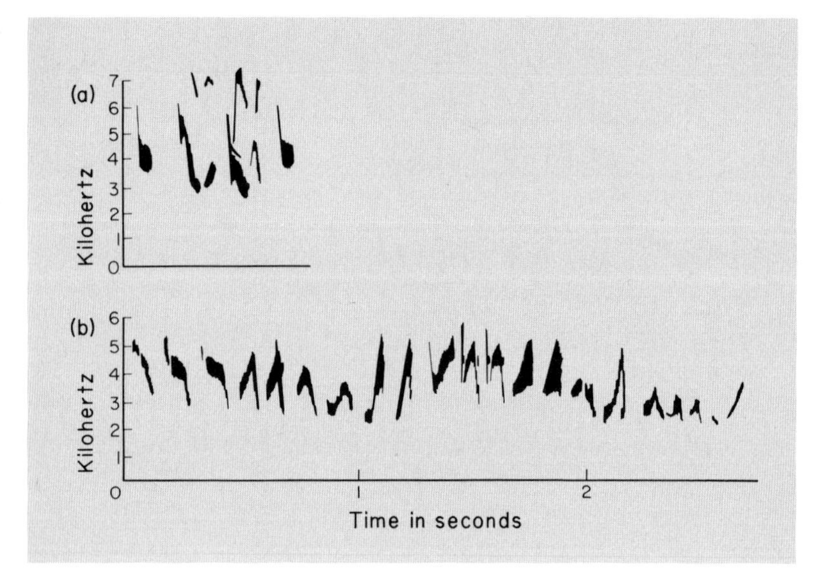

Fig. 7.8 (**a**) Song of the chiffchaff, *Phylloscopus collybita*. This has four phrases, *tsip tsap tsap tsip*. Note the contrasts in general pattern, in duration, and in the detailed structure of the notes. (**b**) Song of the willow-warbler, *P. trochilus*. (From P. Marler,[231] 1957, Fig. 2, *Behaviour* **11**, 13.)

investigated. Good field naturalists cannot always distinguish the song of the reed-warbler (*Acrocephalus scirpaceus*) from that of the sedge-warbler (*A. schoenobanus*) or that of the blackcap from that of the garden-warbler. The habitats of the two members of each of these congeneric pairs are, so far as man can see, exactly the same, and they may be found breeding and singing within a few yards of each other in an environment that is as near uniform as it can be. If each species-pair is derived from a common ancestor the song must have changed little while speciation has been going on. One can perhaps see a further stage in the same process in the chiffchaff and willow-warbler, two congeneric species with distinct preferences in habitat though with some overlap. The birds are not distinguishable by sight except in the hand, but their songs are normally easily recognizable (Fig. 7.8). Occasionally a bird may be heard singing a song which is a mixture of the two, usually two or three of the double notes of the chiffchaff followed by the descant of the willow-warbler. No such bird seems to have been investigated to find out which species it resembles in other ways, or whether (a remote possibility) it is a hybrid.

Confusion can also occur with birds that mimic others. There can be no doubt of this when a bird such as the red-backed shrike, that has an undistinguished song, interpolates an almost perfect rendering of a chaffinch, or when starlings produce imitations of non-vocal sounds. Other examples are less easy to judge; when a redstart (*Phoenicurus phoenicurus*) sings phrases that resemble those of the closely-related nightingale it may be imitating something that it has heard, or it may be producing a genetic variant of its own song. Marsh-warblers (*Acrocephalus palustris*) have been recorded as mimicking 39 other species, and the North American mocking-bird (*Mimus polyglottus*) 51.

There is some evidence that paired birds can distinguish the song of their mate from all others, and occasionally men can recognize individual singers. No doubt they could do better with practice.

7.13 Song dialects

In many species, just as there are geographical differences in colour or size, so are there in the song, sometimes, but not always, corresponding to the subspecies determined on the former characters. Such differences in song are perhaps best shown, or at least most easily studied, in songs of relatively stereotyped pattern.

All normal chaffinch songs are built up musically of a subject and a cadence, but these may be used in different ways. In the commonest, the subject, of five to eight quavers of the same pitch, is announced and then repeated more loudly and at a slightly lower pitch. The cadence, of three to five notes, follows even more loudly. This is shown in the sound spectrogram in Figure 7.9a, which shows also that the subject is preceded by grace-notes, but these are not always present. In variants of the song the number of notes in subject or cadence may be altered, the subject may be given three times or the cadence twice or three times, the number of grace-notes may be extended, and there may be variations in pitch or quality. A common type of song in the European chaffinch, which is a different subspecies from that of Britain, is shown in Figure 7.9b. The extreme form in European birds is probably found in those of Finland, which give out a separate clicking sound about half a second after the close of the cadence. It is easy to fail to realize that this is part of the song.

An individual bird may sing more than one song, even as many as six,

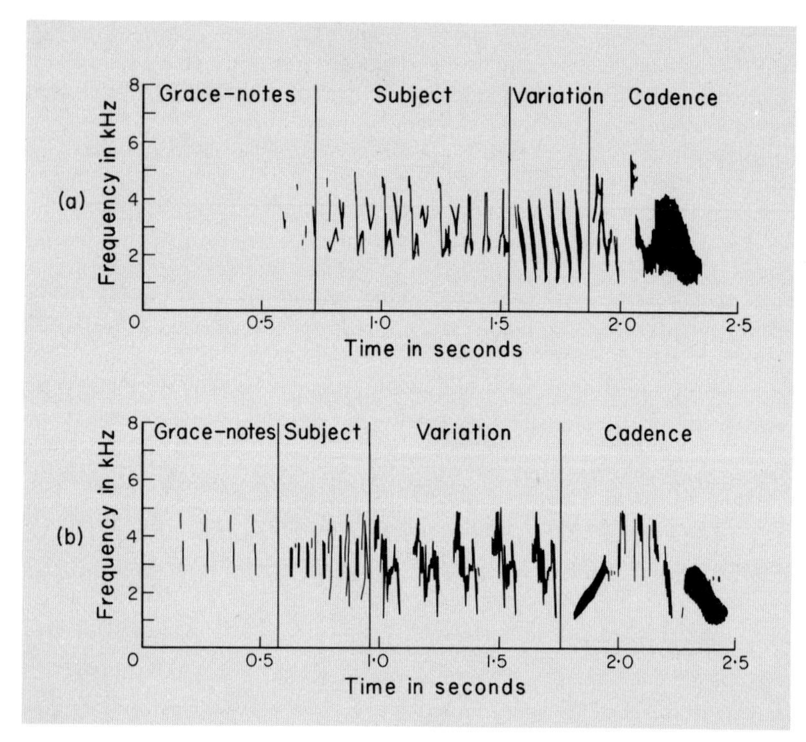

Fig. 7.9 (**a**) Typical full song of the chaffinch, *Fringilla coelebs gengleri* recorded at Madingley, Cambridge. (**b**) Typical full song of the chaffinch, *F.c.coelebs*, recorded in Denmark. (After Thorpe,[362] 1961, *Bird Song*, Figures 21 and 22, University Press, Cambridge.)

but uses mainly one. As will be described below, these individual variants are learnt, not inherited, so that it seems probable that the main geographical dialects are learnt also. No investigations of the more extreme types, such as the Finnish songs, seem to have been made, but it seems likely that they are genetic.

Comparable geographical dialects are found in the much simpler song of the Carolina chickadee (*Parus carolinensis*), which has only a few notes, varying in number, pitch and degree of slurring.[381]

7.14 Learning to sing

The songs of birds are, with few and partial exceptions, characteristic of the species. Are they, therefore, innate specific characters, or are they, like English and French for men, learnt afresh by each individual?

The only way of testing these possibilities is by rearing birds in isolation from the egg, or at least from a very early age. Such experiments have now been done on a number of species, and the results are complicated and perhaps surprising. There seem to be two main types of development of song in the individual, but not enough work has been done for us to say which is the commoner.

In the first, illustrated by the yellowhammer, corn bunting[363] and song sparrow,[232] the song of isolated birds is exactly like that of normal wild ones so that it seems to be completely innate.

In other species, the song has both an innate and a learnt component. If chaffinches are taken from the nest during the first few days of life, and reared in auditory isolation, they begin singing, like their relatives in the wild, in the following spring, but the song that they produce is much simpler than the normal one (Fig. 7.10). It is of about the right length and

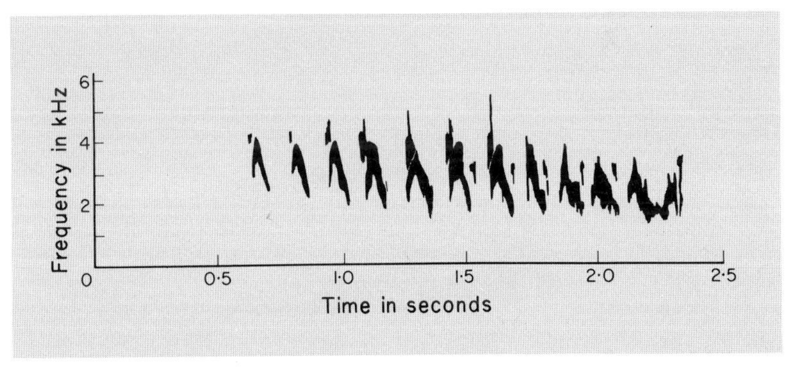

Fig. 7.10 Song of the chaffinch, hand-reared in auditory isolation. (From Thorpe,[362] 1961, *Bird Song*, Fig. 37, University Press, Cambridge.)

the notes are of about the right timbre, but there is no division into subject and cadence, and the pitch is slightly wrong. If two or more such isolated birds are allowed to sing together they improve slightly, but still do not sing perfect songs. If however birds are caught in their first September and then isolated, they sing nearly perfect songs in the following spring, and completely perfect ones if at this time they are allowed to hear other chaffinches singing.

Wild chaffinches cease singing soon after the young are hatched, and rarely sing in the autumn. It seems therefore that young birds still in the nest must learn and remember their parents' or neighbours' songs, although they do not use them until eight or nine months later, at which time they can learn details of other songs that they hear. Since birds tend to return to the neighbourhood of their birthplace to nest, it is presumably in this way that local song dialects are built up. In artificial situations isolates can be trained to imitate other songs of the same general timbre, for example that of the tree-pipit (*Anthus trivialis*). The mimicked components of some birds' songs may be acquired in this way.

An interesting use of this power of learning the song of other species is found in the combassous (*Hypochera*) of west Africa.[283] They are small black birds, and occur in a number of forms, very difficult to distinguish

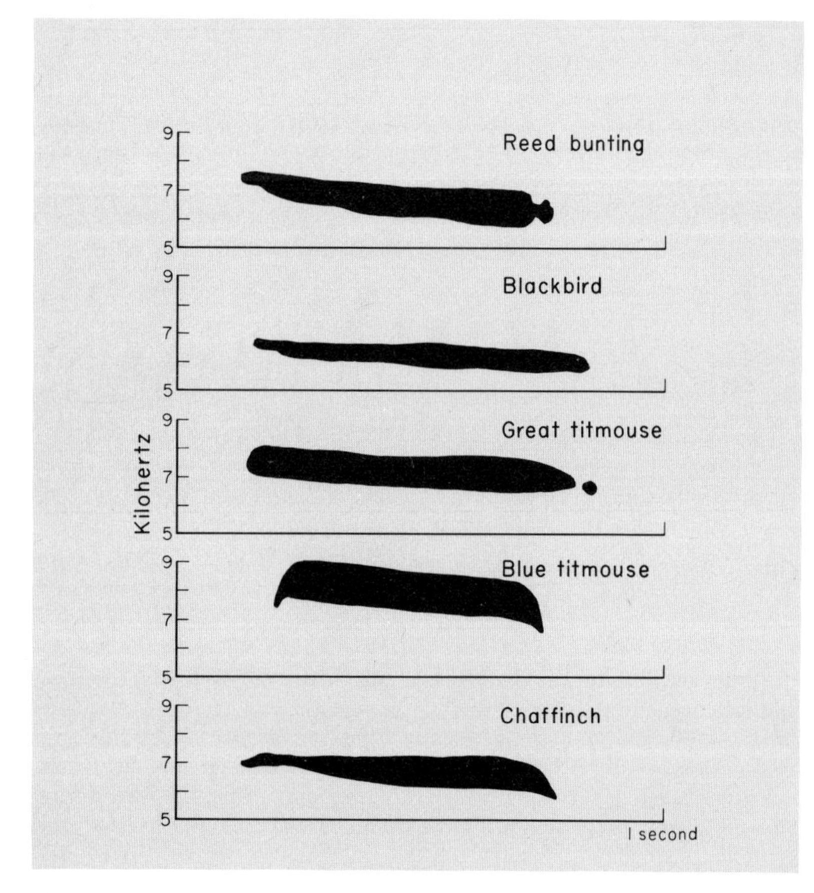

Fig. 7.11 The calls of five different species given when a hawk flies over. Each has a relatively narrow frequency range, somewhat over-emphasized in this illustration as a result of using wide-band pass filters. They all sound like a high, thin whistle, and are difficult to locate. (From P. Marler,[231] 1957, *Behaviour* **11**, in Bell, 1959, *Darwin's Biological Work*, Fig. 17, University Press, Cambridge.)

from one another. They are brood parasites, and each form is more or less restricted to one species of fire-finch (*Lagonostricta*) as host, whose song it mimics. Experiments on isolation have not been done, but presumably the song is learnt in the nest, as by chaffinches. Since the females choose their mates by song there is good reproductive isolation, and the song would seem to have been the main agent in speciation. It is indeed more constant than the restriction of one colour form to one host.

Other species that acquire their songs in a similar way to the chaffinch are the blackbird and white-crowned sparrow.[232] Since females of the last species injected with testosterone sing the dialect of the area where they were collected, they too must be able to learn and remember the parental song, even though they never normally use it.[191]

In birds that need to hear the parental song before they can sing, and in some of those in which song is innate, some practice is necessary before they can sing properly, and they need to hear their own voice. If they are deafened before they have acquired their song (for example in their first autumn) they do not achieve perfect songs. This has been demonstrated for, amongst others, the American song-sparrow which does not need to hear its parents, and for the chaffinch which does.[276]

7.2 CALL-NOTES

The other sounds made by birds, which are called collectively call-notes, seem to be entirely innate. They are often less specifically distinct than songs; some of the contact notes of the tits, for example, used to keep the members of a flock together, are very similar, and as the flocks consist of several species the advantages are obvious. Alarm calls also may be similar even in unrelated species[231] (Fig. 7.11). This would appear to be a form of vocal Müllerian mimicry. The figure illustrates also that such calls are of high pitch and usually begin and end gradually, so that they are difficult to locate. It is clearly important that a sound that gives warning of the presence of a predator should not also betray the whereabouts of the caller.

Many, but not all, bird-songs and calls have ultrasonic components,[364] but since all the evidence is that most birds are deaf to these high frequencies they are probably functionless.

7.3 TERRITORY

The concept of territory was first brought to the notice of ornithologists (and zoologists) over half a century ago by Eliot Howard,[168] although the word had previously been used in the same sense by Moffat in 1903.[253] It has been defined in various ways, the shortest and most comprehensive being 'any defended area'. Most discussions of the matter, however, deal with something much more limited than this, and we shall discuss in this chapter only areas that are large in relation to the size of the bird, and that are defended in connection with reproduction, or are accompanied by other reproductive phenomena.

Territory is most typical in those species, such as many passerines (finches, tits, larks, wagtails, some thrushes) that live in flocks during the winter but form pairs that nest apart from others in the spring. The selection of a nest-site by such birds has already been described; the nest-site is surrounded by a more or less extended area which is defended against other members of the species by the cock, and sometimes by both sexes. This area is the territory. In some species, especially hole-nesters, the nest-site is chosen first and the territory is formed round it, in others the territory is occupied first and the site of the nest is later chosen within it; this happens in those birds, probably the majority, where any area of suitable habitat and appropriate size is likely to contain several possible nest-sites. The cock wren (*Troglodytes troglodytes*) may begin to build as many as 12 nests in his territory, and his average number is six.[17] Usually the territory is established by the male alone, often, in migrants, before the females have arrived. In birds that sing from trees or bushes it contains one or more song-posts, more or less conspicuous perches from which the holder habitually sings; they are often distributed round the perimeter of the territory rather than placed at the centre. Song-posts usually stand out in some way from their surroundings, as is well shown by garden-nesting blackbirds which often sing from the edge of the house roof, or starlings which prefer the more elevated aerials. The tree-pipit usually begins its song and song-flight from a tree 30 or more feet tall, but on new plantations may be satisfied by one that is only two or three feet high but is equally conspicuous in relation to its surroundings.

A peculiar demonstration of territory is the roding of woodcock (*Scolopax rusticola*). The cock, and occasionally the hen, flies at dusk and dawn 10 or a dozen times round a circuit of about one to three miles, which is usually a long ellipse but may be a figure of eight. The flight is slow, usually just above the tops of the trees, and every so often the bird checks its flight and calls. Intruding woodcocks are driven off.

The size of territories varies so much, both between species and within species, that without series of figures with proper statistical treatment, which have not been published, few meaningful statements can be made about it; perhaps the best is that it is usually of the order (in the strict sense) of one acre, or one hectare. Some limits that have been recorded (all, except for the wheatear, on small numbers) are shown in Table 7.1. Species such as dippers (*Cinclus cinclus*) and kingfishers (*Alcedo atthis*), that are restricted to streams and their banks, have territories that are linear rather than areal. So too may birds that live primarily in woodland

Table 7.1 Size of territories.

Species	Acres	Hectares
Willow-warbler[241a]	0·16 – 1·2	0·065 – 0·49
Chaffinch[230]	0·25 – 2·9	0·1 – 1·2
Robin[194]	0·4 – *c.* 2·0	0·16 – *c.* 0·8
Blue tit[127]	0·4 – 2·0	0·16 – 0·8
Great tit[127]	0·6 – 4·2	0·24 – 1·7
Marsh tit[127]	1 – 16	0·4 – 6·5
Willow tit[127]	to 65	to 26
Wren[17]	*c.* 1 – *c.* 7	*c.* 0·4 – *c.* 2·8
Wheatear[66]	1·2 – 8·1	0·49 – 3·3
Red-backed shrike[91]	<1·4 – 8·3	<0·5 – 3·4
Corn bunting[16]	1·7 – 2·7	0·7 – 1·1
Tawny owl[334]	*c.* 10 – *c.* 30	*c.* 4 – *c.* 12

or scrub but also make use of hedges; the yellowhammer (*Emberiza citrinella*), for example, may occupy a length of hedge from less than 60 m to more than 130 m.[16]

Territory is defended primarily against members of the same species, but sometimes, as in the hawfinch[263] and red-backed shrike,[91] others are driven off as well. Since the general density of birds in most habitats is much more than one pair to an acre this interspecific defence must be relatively rare. (Defence of the immediate surroundings of the nest against intruders in general, which would be territorial according to the wide definition quoted above, is common.)

A few species hold territory in the winter. Robins do so if they do not migrate, and the resident females of this bird take up territories in autumn distinct from those of their mates. In gardens, both in England[330] and Denmark,[210] many cock blackbirds hold their territories throughout the year, relaxing their defence only during the moult in August, and their mates stay with them; in Danish gardens some solitary hens hold territories. In woods neither sex does so after the breeding season. The difference could be due to the better feeding in gardens, or to the higher density. Some resident British starlings have been described as holding territory in autumn, while others, and all the migrant visitors around them, are in flocks, but the areas defended are little more than their song-posts.[259]

In South Africa at least 20 per cent of the resident terrestrial birds (excluding brood parasites) hold territory as pairs throughout the year, and probably for life. The marsh tit in England, and the plain tit (*Parus inornatus*) in America, are similar.[305]

The setting-up of territory in spring coincides with the onset of song, the courtship of the other sex, and the swelling of the gonads and increase of sex hormones. It may be gradual, and, like these other things, affected by the weather. Many male birds, buntings for example, leave the flock in sunny weather, sing in their territory for an hour or two, and then return. Although little experimental work has been done, we can be confident that normal territory-holding is part of the sunlight-pituitary-endocrine cycle described in Chapter 5; cock grouse (*Lagopus lagopus*) can be induced to occupy territory by implantation of testosterone, and inhibited from doing so by oestrogen.[16a] Although the evidence is less clear, it seems likely that the holding of winter territory is a development of the same cycle. It is often accompanied by song, even in the hen when she is territorial, and both sexes of the starlings that hold territory in the autumn have a higher than normal production of testosterone. Garden blackbirds leave their territories in cold weather.

7.31 The function of territory

The occurrence of territory, and its connection with reproduction, are clear, but the means by which it has been produced by natural selection, and so its biological function, are doubtful and have been much debated. Its effect must be to space birds throughout the habitat, so that if more than one potential nest-site is present in the territory the effect must be to limit population. If, however, nest-sites are restricted and fully occupied, territory cannot directly affect population and merely achieves privacy for the occupants.

The figures in Table 7.1 show that the size of territories varies widely, and experiments of two sorts have been done to see how populations can be affected by variations in the conditions. In one set of experiments nearly all the territory-holding males were removed from an area of spruce woodland in North America.[160, 346] As new males came in they too were shot, until by the end of the season the number of killed was nearly three times that present at the beginning. This is taken to mean that the surplus birds were previously kept out by those already holding territory, so that the breeding population was limited in this way. The same thing is shown less destructively by the general observation that whenever a new area of a given type of habitat is provided, as by the plantation of woodland, it is rapidly populated by the appropriate birds, without any reduction in their numbers in adjacent pre-existing habitats of the same sort.

More precise observations, though for a limited range of species, have been made by the use of nest-boxes. Some species, especially hole-nesters, readily use them, and may prefer them to natural sites. By their use populations of pied flycatchers have been induced, without any other change in the habitat, in areas where previously there were none. It is easy to distinguish between the limiting effects of nest-sites and of territory by increasing the density of the boxes.

Experiments in the Netherlands[190] suggested that in a heavily-boxed but otherwise unfavourable pinewood habitat the densities of blue tits, coal tits and great tits (*Parus major*) varied widely, but did not rise above about 50 per cent occupancy of the boxes by the three species. In nearby mixed woods, apparently more favourable, with natural nest-sites but fewer boxes, the densities were higher but more nearly constant. The great tit in particular fluctuated little over the 12 years 1941–52. In both habitats something other than the number of nest-sites must have been limiting the population, and the most likely factor is territory. The same thing is shown by pied flycatchers in the Forest of Dean.[56] These birds were previously unknown there, but appeared immediately following the the provision of nest-boxes in 1942. As the boxes were increased to about 200 on 810 acres (320 hectares) the flycatchers increased too, but by 1951, when there were 100 pairs, they seemed to have reached a maximum, since there were many unoccupied boxes (some were used by other species), and the numbers have fallen since then although the number of boxes has remained about the same. At the maximum density the mean size of territory is eight acres (3·2 hectares) (Table 7.2).

By contrast, in Finland successive increases in the number of nest-boxes continued to increase the numbers of pied flycatchers[143] (Fig. 7.12). A limiting density appears to be reached only when the boxes are 25 m or less apart.[368]

Although birds do, when necessary, defend their territory by active fighting, as well as by challenge through display and song, there is no evidence that defence as such is the regulating mechanism. Persistent males may carve out for themselves enough space for breeding in a habitat already fully occupied in the sense that there is no part of it that is not in some bird's territory. Possibly the mere presence of a high density of holders, which will be manifested by a high output of song, makes an area unattractive to a newcomer, which is then likely to go elsewhere, and may or may not find a place where it can breed.

The most obvious value in territory would be that of ensuring an adequate food supply for the pair and their brood. Clearly there must be a limit of density beyond which an area cannot supply enough, and it is noticeable that the species which show the strongest territorial behaviour are those such as tits, finches and thrushes that collect their food in some definite places near the nest. Birds that feed on the wing, like swallows, are not strongly territorial, and even the pied flycatcher, which feeds on passing insects, does not defend the perimeter of its territory very strongly (though it is pugnacious around the nest-site). There is no consistent evidence that the size of territory is inversely proportional to the amount

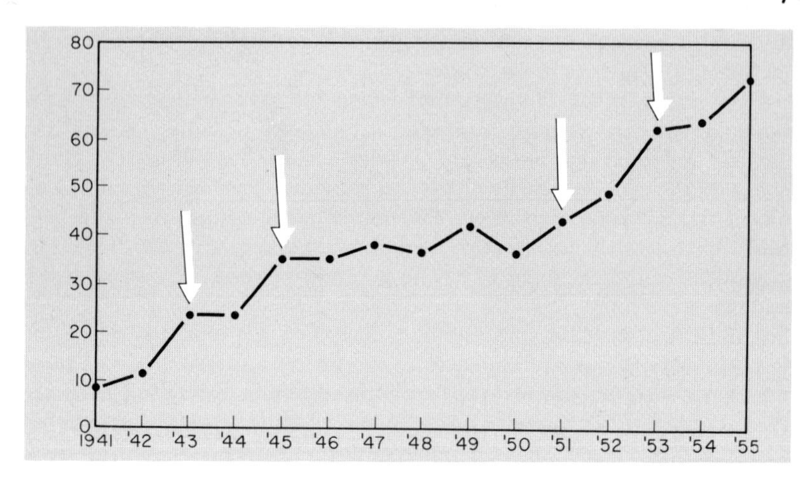

Fig. 7.12 Number of nest-boxes where clutches were laid by pied flycatchers at Lemsjöholm, 1941–55. Arrows indicate years in which these numbers were considerably increased. (From L. von Haartman,[143] 1956, Fig. 1, *Ibis* **98**, 461.)

of food available, but the tits in Dutch woods mentioned above show a rough inverse correspondence. Some species, such as reed-warblers and sedge-warblers, feed mainly outside their territories, and others, like the robin, allow other individuals to feed in theirs. In the pied flycatcher there is no correlation between breeding success and density. The value of territory in controlling food supply, if it is present, is indirect, or perhaps it is only present at a brief period during the growth of the young, or in occasional poor years.

In most species the territory not only surrounds the nest, but is also the site of other sexual activities, especially courtship and copulation; some, such as gulls and the lek birds, defend areas used for these purposes but not an area round the nest. Privacy and lack of disturbance possibly lead to more successful pairing, and so may have survival value. Other suggestions, such as that territory reduces disease or predation, seem to be pure speculation.

The dominant view at the present time seems to be that territory has more than one function, and that its value differs from bird to bird. If this is true, it follows that the same type of behaviour has been produced in different species by natural selection acting through different advantages. This is not impossible, but is not an economical hypothesis, and seems unlikely for birds of fairly close common ancestry living in similar environments.

Table 7.2 Numbers of pied flycatchers in Forest of Dean.[56]

Year	1942	1943	1944	1945	1946	1947
No. of nest-boxes	84	91	91	91	91	110
Boxes occupied by pied flycatchers	15	33	35	34	37	54
Year	1948	1949	1950	1951	1952	1953
No. of nest-boxes	144	205	258	(270)	275	(260)
Boxes occupied by pied flycatchers	56	67	87	100	98	85
Year	1954	1955	1956	1957	1958	1959
No. of nest-boxes	(255)	(255)	(255)	(255)	(240)	(240)
Boxes occupied by pied flycatchers	76	67	60	54	71	71
Year	1960	1961	1962	1963	1964	
No. of nest-boxes	(240)	251	246	245	238	
Boxes occupied by pied flycatchers.	58	62	59	58	57	

Notes. Data for 1942–7 not of comparable accuracy to those from 1948 onwards. Totals of nest-boxes available bracketed in some years when one or two were damaged early in season and not always recorded.

7.4 MIGRATION

Migration is the to-and-fro movement of a population of animals between one area, where they breed, and another, where they do not. Often, and perhaps usually, more than half the year may be spent in the non-breeding area (which for convenience, though not entirely accurately, may be called the winter quarters) or on the way to and from it. Birds are not the only animals that migrate, but because of their mobility they show the behaviour to an extent unparalleled in any other group. Few species are wholly resident on one patch of ground, and although some merely scatter after breeding, so that there is no migration as we have just defined it, others, though nominally 'resident' in a country according to the lists, make small regular movements; grouse from hill to valley and chaffinches from oakwoods to farmlands for example. Many species that may be found throughout the year in Great Britain leave northern Europe entirely after the breeding season. Even in the tropics there is movement of this sort; birds that breed in the savannah of northern Nigeria, for example, may move south into the rain forest at the beginning of the dry season, and some species mix thus with members of the same species resident in the forest all the year.

Typical bird migration in the northern hemisphere is a movement between a northern breeding ground and winter quarters further to the

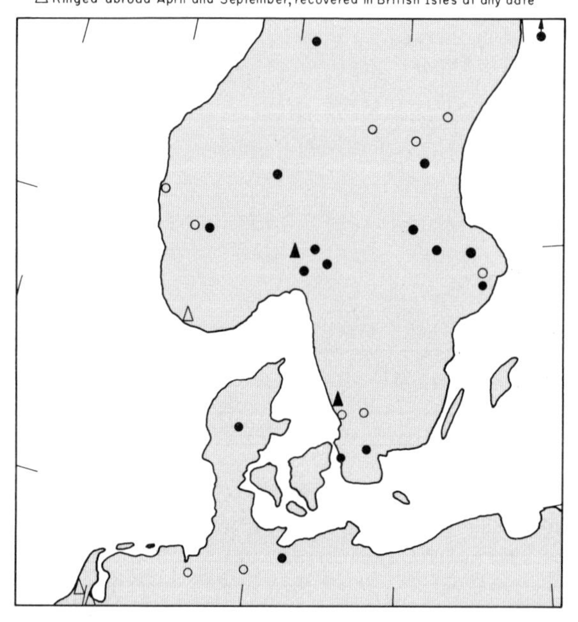

Fig. 7.13 Breeding range of chaffinches that spend the winter in the British Isles. (From Goodacre,[132] 1959, Map 1, *Bird Study* **6**, 103.)

south. It may be within the temperate regions, as from northern Europe to Great Britain or southern Europe (e.g. the chaffinch,[132] Fig. 7.13) or within North America (e.g. the white-throated sparrow, *Zonotrichia albicollis*, Fig. 7.14). It may be from temperate or subarctic parts to warm temperate or tropical Africa or America (e.g. the blackcap, Fig. 7.15, and the ruby-throated humming-bird, *Archilochus colubris*). Some temperate migrants cross the equator and spend the northern winter in the southern

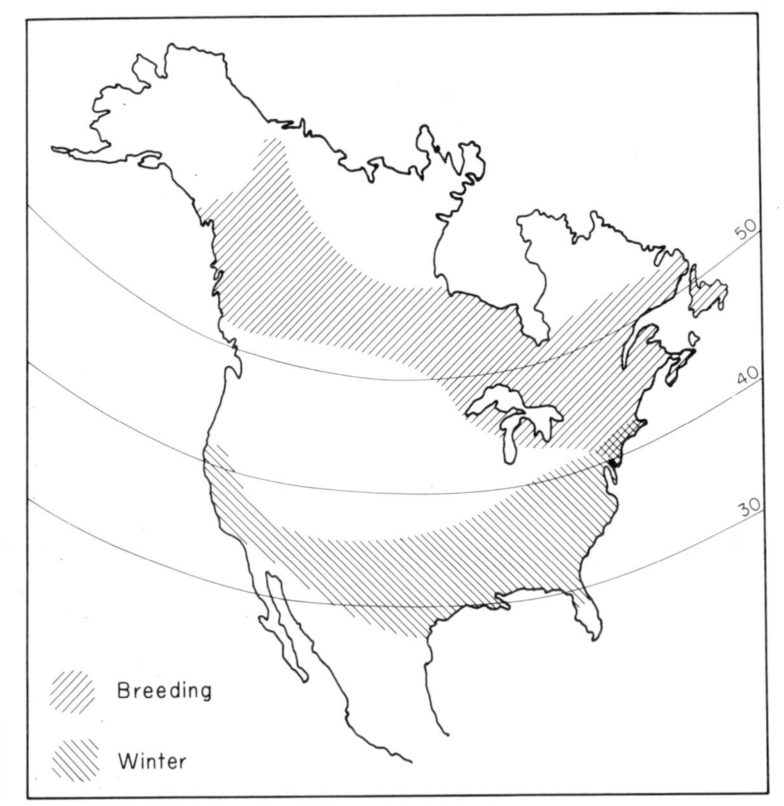

Fig. 7.14 Migration of the white-throated sparrow within North America

summer (e.g. the swallow, *Hirundo rustica*,[73] Fig. 7.16, and the bobolink, *Dolichonyx orizivorus*). Some species, especially waders that nest in the Arctic, fly almost half way round the globe. The American golden plover (*Pluvialis dominica*), for example, moves between Alaska and Patagonia (Fig. 7.17), and 14 species that nest in the Arctic go to New Zealand for the other half of the year.[106]

Several species, such as the starling (Fig. 7.18), move from a breeding area in east Europe or Asia towards the Atlantic coast, so avoiding the continental winter as effectively as if they moved south. Some sea-birds also make migrations that are longitudinal rather than latitudinal; for example sooty terns (*Sterna fuscata*) breed on the Dry Tortugas just north of the Tropic of Cancer in the Gulf of Mexico, and cross the Atlantic 6000 miles to the west coast of Africa.[301]

Migration has been less studied in the southern hemisphere because

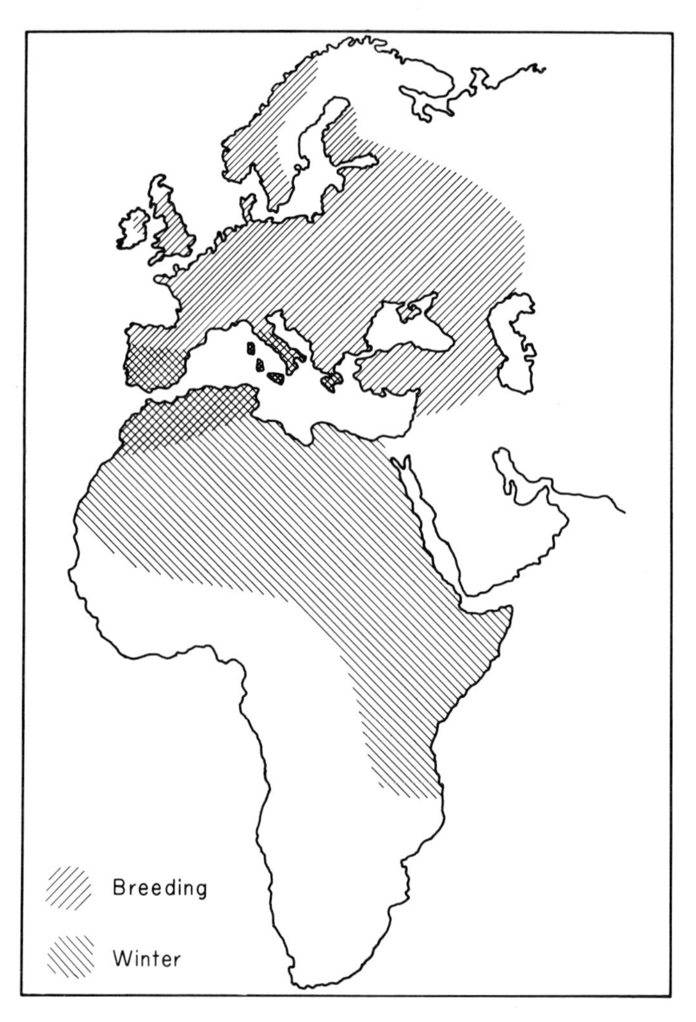

Fig. 7.15 Migration of the blackcap from temperate regions to warm-temperate and tropical Africa. There is some overlap, and a few birds winter further north than shown, for example in Southern England

there are fewer ornithologists there, but there is less land in southern high latitudes than in those of the north, so that there cannot be the same massive movements. Several species however migrate northwards from breeding grounds in South Africa or southern South America.

Four questions must be asked about migration. What advantage does it

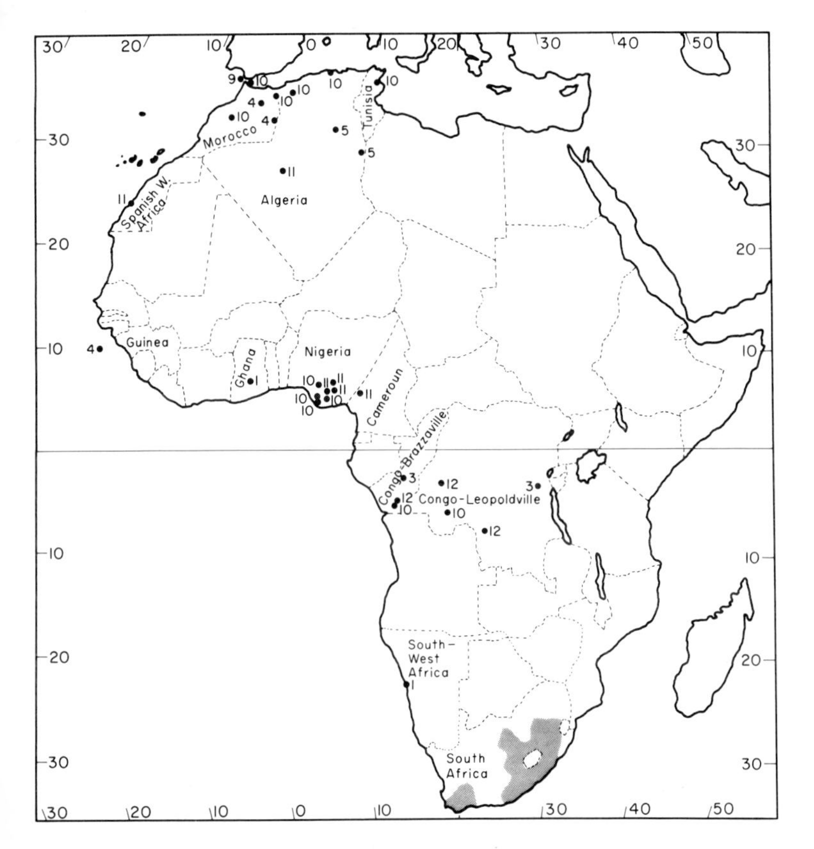

Fig. 7.16 Recoveries of British-ringed swallows in Africa. The figures indicate the month of recovery. The wintering area in S. Africa (54 recoveries) is shown shaded. (From Davis,[73] 1965, Fig. 3, *Bird Study* **12**, 158.)

confer on the birds, so that the instinct has been selected? What induces the movement in the individual? How are the long flights of many species sustained? How do the birds find their way?

7.41 The advantages of migration

A partial answer to the first is that the Arctic, where many migrants breed, is completely unsuitable for bird life except for about three months of the year. The ground is frozen, for some time there is no sunlight, the vegetation (not much used as food by birds anyway) is buried under snow, and there is no active insect life. The birds must therefore leave; what needs to be explained is why they come back.

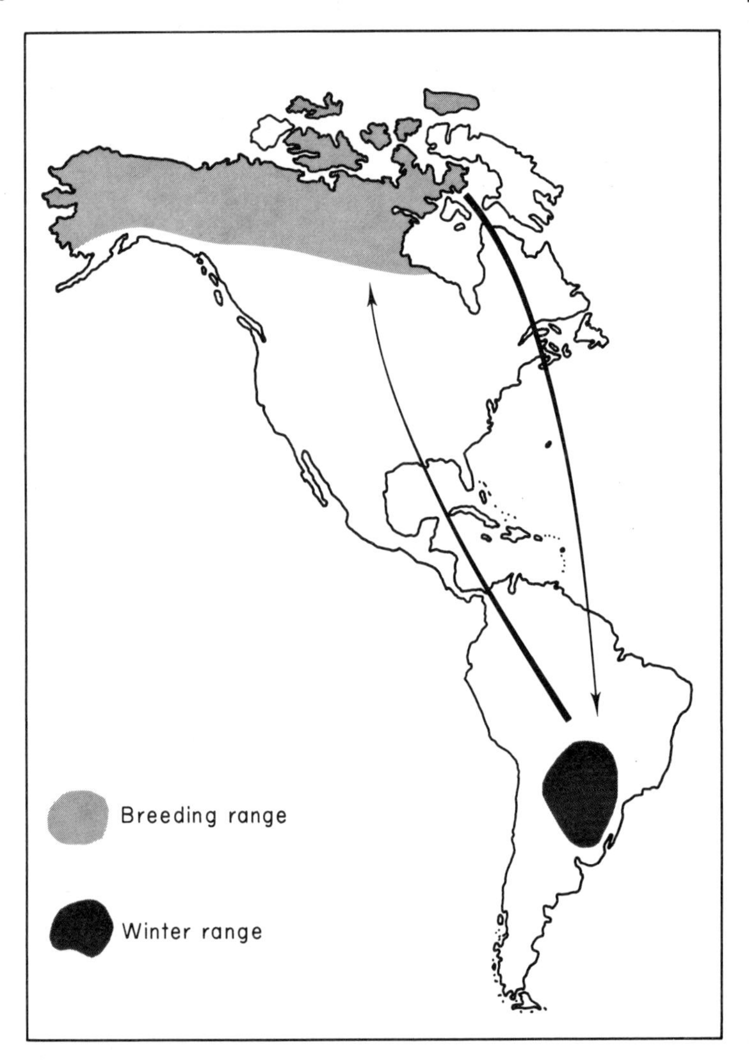

Fig. 7.17 Migration of the American golden plover. The arrows show the approximate tracks of the spring and fall movements

Even in the temperate zone conditions are obviously less favourable in the winter. There is little primary production and there are few flying insects, and the ground is intermittently frozen. Yet the tits (Paridae) manage to spend the winter in Great Britain under these circumstances, with migration only away from the more exposed areas such as the woods

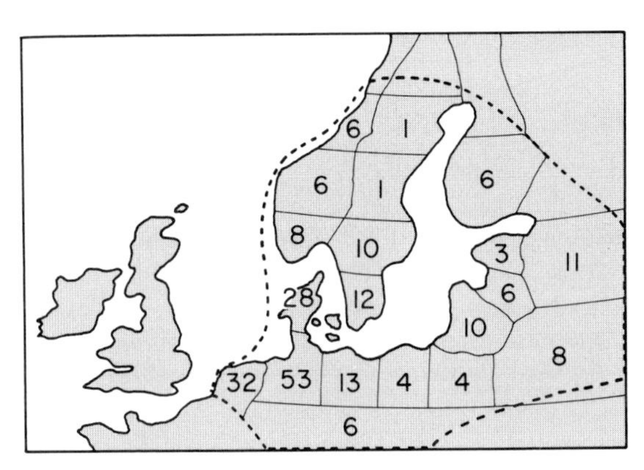

Fig. 7.18 Breeding range of starlings that spend the winter in the British Isles. Numerals indicate the breeding-season recoveries (May–Aug.) of starlings previously ringed during winter (Dec.–Feb.) in the British Isles. (From Goodacre,[132] 1959, Map 1, *Bird Study* **6**, 181.)

on the hills, while the warblers (Sylviidae), which take very similar food, all leave with the exception of the Dartford warbler (*Sylvia undata*), which hangs on precariously in its few breeding places in the southern counties, and occasional blackcaps and chiffchaffs.

The last two species illustrate a somewhat surprising point, that the species that nest further north tend to go further south on migration, for their close relatives the garden-warbler and willow-warbler, which nest alongside them here but extend further to the north both in Scotland and Europe, have seldom or never been recorded in England in the winter. In accordance with this they extend further to the south in Africa. In one season's netting in northern Nigeria, for example, 197 garden-warblers were recorded, but only four blackcaps.[88] The rule, however, is not fixed; the red-spotted bluethroat (*Cyanosylvia suecica suecica*) for example, which breeds exclusively in the far north of Europe except at high altitudes, does not go beyond 10°N in its winter quarters in Africa, while its relative the nightingale, which does not breed further north than Germany, winters in tropical Africa, including places well south of the equator.

The ecological conditions over most of Africa are determined much more by the alternation of wet and dry seasons than by the conventional winter and summer of Europe. For most of southern Africa the migrants arrive about the beginning of the rainy season,[256] and there is then an abundance of insect life far beyond anything in the temperate zone. Even in the dry season there is probably in many places a greater density of invertebrate biomass than in Europe in summer. One may assume, therefore, that there is more than enough food to go round, even though the migrants are competing with a large resident population. Some of the visiting species will be occupying different niches from the residents, but for others there seems to be complete identity of insect prey. Migrant European bee-eaters (*Merops*) and swallows (*Hirundo*) feed in the same way and in the same places as the resident species, and so far as is known take the same sort of insects. Moreau has described the European yellow wagtail (*Motacilla flava*) in east Africa as following the cattle and feeding on the insects they disturb, a niche unoccupied by any local small bird,[256] but in Nigeria its habits and feeding are indistinguishable from those of the local pied wagtail (*M. aguimp*), both species feeding on short grassland as the wagtails do in England.

It may be accepted in general that migrants get more food, and better conditions, in their winter quarters than they would if they stayed at home. But why do they come back? The short answer is that there is food to be exploited. If it is generally true, as it appears to be, that at the beginning of each breeding season there are, in most species, more pairs than there are nest-sites or territories, then any extension of appropriate habitat will be occupied, even if it is suitable for part of the year only. This, with the addition of attachment to the nest-site, will account for northward spread into relatively unfavourable areas; it has been recorded as going on during the climatic amelioration of the last 150 years. Several species (tree-pipit, chaffinch, blue tit, wood-warbler) have reached the woods of the northwest highlands of Scotland only during the last 100 years,[406] and Iceland gained nine breeding species in the first half of this century.[141] In many species, though not all, there is an increase in the average size of the clutch as one goes north, and, whatever the reasons for this,[101] the larger broods (assuming that there is no countervailing increased mortality of the nestlings) will be able to compensate for any increased mortality brought about by the hazards of the migrational journey.

The explanation just given would account for the origin of a migratory population on the northern or unfavourable edge of a species' range. This stage has been reached by species such as the chaffinch, whose Scandinavian populations migrate while those in mid-Europe do not. Competition with the resident birds might tend to drive the migrants beyond them to winter quarters where the species is absent, so accounting for the 'further north, further south' rule. If new migrant species are to be brought into existence reproductive isolation of the populations must occur. The migratory habit itself might produce this, although one would expect a broad zone of overlap and mixture. Such occurs between the distributions of the hooded crow (*Corvus corone cornix*) and carrion crow (*C. corone corone*) which are separate species, or subspecies of *C. corone*, at the reader's choice. The carrion crow, in England and western Europe, is mostly sedentary, while the hooded crow, which occupies most of continental Eurasia, is largely migratory. If they are not yet separate species they are well on the way to becoming so.

7.42 The stimulus for migration

Whatever may be the advantages that have led to the selection of the migratory instinct, there must be other proximate factors which act on each individual to induce it to move at the appropriate season. In a general way migration is part of the sexual cycle. Birds begin to move northward as their gonads begin to swell, and by the time that they arrive on the breeding grounds the glands are active. It does not follow that the sex hormones are the cause of migration, and a number of authors have argued that they are not.

Actual migration is impossible to follow while experiments are being carried on, but in many species the movement is preceded by a premigratory restlessness[386] (well shown by the behaviour of swallows in autumn), and by the deposition of fat,[187] both of which can be studied in caged birds that are subjected to various treatments. The results of experiments have not been entirely clear or consistent, but it is possible to induce both restlessness (often called by its German name of 'zugunruhe') and deposition of fat by manipulating the length of day, or photoperiod.[214, 388] This is known to affect the pituitary (Chapter 5), which produces gonadotropins, so the connection through the gonad seems probable. Castrates certainly sometimes migrate, or show restlessness, but they might have been socially stimulated by intact birds, or the necessary steroid hormones might have been produced by the adrenals. In some experiments castration has been claimed to reduce restlessness.[389] The accumulation of fat, in both caged and wild birds, is sudden and rapid, especially in spring, and takes about 10 days in finches.[186] It is not the cause of the restlessness, for in a number of species the latter has been induced in birds that have been prevented from putting on weight by limiting their food supply.[215] Restlessness might, on the other hand, cause fattening, since it might lead to over-eating, which has been shown to accompany fat deposition in the chaffinch and the white-crowned sparrow.[182]

In some experiments restlessness and fattening have been induced by rising temperature. Low temperatures and other unfavourable conditions inhibit them even when the light ration is suitable. For the many species that winter in the tropics, where changes in length of day are very small, and different on the two sides of the equator, it seems that some stimulus other than light must act.

There are specific physiological differences between birds that migrate and those that do not. Non-migratory house-sparrows in the United

States do not become restless or put on weight, while the closely-related American tree-sparrow (*Spizella arborea*) does.[182] The non-migratory yellowhammer can endure lower temperatures than the ortolan (*Emberiza hortolana*) which breeds alongside it in Finland but migrates south of the Sahara.[380] In a series of American species the amount of fat accumulated is greater in those that migrate over ecological barriers or by night than those that fly only over suitable feeding grounds or by day.[187]

Whatever the exact stimulus, the urge to migrate is brought on by something that is part of the annual seasonal breeding cycle; it is not necessarily the same in all species and may be helped by an innate rhythm. When the bird is in a condition to migrate, something must start it moving, especially when, after the breeding season, the movement is a sudden leaving of the region of the nest; swallows and house-martins may be around the house one day in September, and the next day all are gone and are not seen again until the following April. Little is known of this, but high atmospheric pressure and clear skies are often associated with departure from the breeding grounds; they will give constant good weather conditions, warm days and cold nights. The advantage of these for long flights is obvious, but which of the factors is the trigger is unknown.

During the migration weather conditions are also important, and cold and storms hold up the homeward movement. That of the willow-warbler[332] and some other species over Europe is closely related to temperature (Fig. 7.19) and a similar relationship has been shown for some American species[209] Conversely, a sudden lowering of temperature in winter may bring migrants such as redwings (*Turdus musicus*) to parts of western and southern England which they have not previously reached.

7.43 The physiology of migration

As long as a bird is passing over land other than desert its migratory movement need bring no difficulty, for it can feed as it goes. Swallows, which are some of the easiest birds to observe in migration, may often be seen in September flying steadily south, either in a straight line or with circling, and they can catch insects in this way as easily as usual. Warblers make their way slowly south, sometimes stopping to feed, and may even sing for a day to two in one spot. All long-distance palaearctic migrants, however, must cross the seas, or the desert belt stretching from north Africa to central Africa, or both, and this means that they must fly for some hundreds of miles without food or drink. North American birds may similarly cross the Gulf of Mexico.

It is here that the importance of the premigratory deposition of fat comes in. From a usual non-migrating level of about five per cent of wet weight the proportion of fat rises in many species to about 30 per cent, and in a few to over 40 per cent.[279] The fat is deposited in all organs except the heart, but the bulk of it is subcutaneous. The energy needed in flight is difficult to estimate, and good values are known only for a few humming-birds, but on the most pessimistic assumptions migrating birds carry enough stored fat to keep them going for some hundreds of miles. Nevertheless, on long flights, such as crossing the Sahara, the North Sea or the Gulf of Mexico, there must often be little reserve, especially if there are

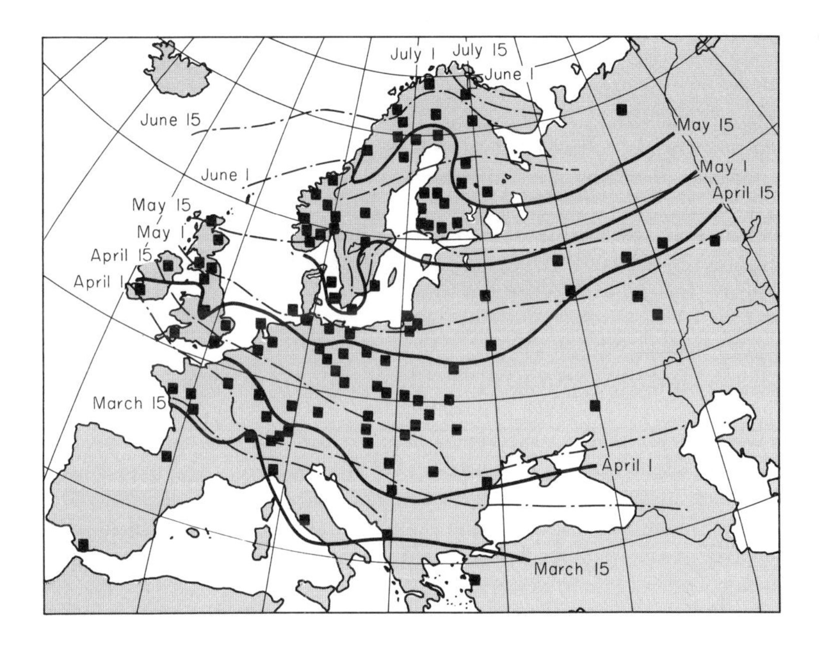

Fig. 7.19 Map showing the rate of spread of the willow-warbler (*Phylloscopus t. trochilus*) over Europe during the spring. The continuous black lines, dated at the righthand ends, are isochronal lines showing the stage achieved every fortnight. The dotted lines, dated at the lefthand ends, show the movement of the 48°F isotherm for comparison. Each black square represents a place for which an average arrival date was worked out from local data. (From Southern,[332] 1938, *Br. Birds* **32**, 205.)

adverse winds, and birds arriving under these conditions are often observed to be exhausted. The physiological advantage of fat over glycogen as a storage product is obvious, since it produces much more energy per unit weight. It also produces more water, and that too is an advantage since with the necessary high rate of metabolism and high rate of breathing there must be much evaporation from the lungs.

7.44 Navigation

The problem of how migrating birds find their way has not yet been solved. It is often confused with the somewhat different one of homing, but the distinction can be made clear by a human comparison. In the days of trench warfare an elementary exercise in the training of an officer was to send him out on a featureless plain in the dark and tell him to find his way back. He could do so by means of a compass, counting his paces, and simple trigonometry. This is homing, and the bird (or any other vertebrate) could in principle achieve it by means of the semicircular canals (which can measure the angle of turn) and some means of determining distance, whether by counting paces or wing-beats or by using time as the measure when speed is constant. There are plenty of examples of homing by mammals, and by non-migratory birds such as pigeons, and although there is some evidence that the birds at least use the sun to help them, such outside clues are not in principle necessary. Experiments on conditioned reflexes show that pigeons can detect an angular displacement from the direction of a light of about 3°; whether they do this by sight alone or by sight plus the inner ear their sensitivity is good enough to account for most homing.[222]

If a mariner were instructed to sail to South Africa, where he had never been before, with no maps and no knowledge of its latitude and longitude, he would be unlikely to succeed. Yet this is the problem that is solved by those species of bird in which the parents leave before the young of the year, such as the cuckoo. This is not homing; although they may have common elements, homing and migration are different things. Even where the young accompany their parents the birds would have to remember a journey made six months before.

Common observation of day-flying migrants such as swallows shows that they follow more or less constant compass bearings, which have been called standard directions and preferred headings. The routes are often deflected by topographical features such as mountains and valleys, and the headings are not necessarily the same throughout a bird's journey; many migrants, for example, fly southwest or southeast through Europe, so avoiding the wide crossing of the Mediterranean, and then change their direction to south. In recent years radar[93] has enabled us to follow for short distances the paths of flocks of night migrants (the species of which can only be guessed, although the general type of bird may be fairly clear), and here also there is some constancy of direction.

Two questions arise: are the birds only following an innate direction or are they using local landmarks? And how far can they compensate for lateral displacement? All airborne creatures are, of physical necessity, subject to drift. The velocity of the wind must be added vectorially to their own velocity to give a resultant which will determine their track over the land. Radar observations both in Britain and in America suggest that night migrants (probably small passerines and waders) maintain a constant track over land, so that they must be compensating for drift by altering their heading.[152] They do this even on dark nights, but features of the the land (for example the Pennines) may nevertheless be visible.[102]

Landmarks obviously cannot be used over the sea, and although wind-lanes and the direction of rollers could in some circumstances give information to the birds they are too variable to be of general use. Crossing the North Sea (unless it is pure chance, the birds flying at random and being carried east to west or west to east according to the prevailing wind), must depend on a definite heading.

Strong winds divert Scandinavian migrants to the northeast coast of England, and recoveries of ringed birds suggest that these birds, bound for Spain, reach that country in as high proportion as those that have taken the direct southwesterly route and have not crossed the sea. They must have reorientated, but could hardly have done so by recognizing landmarks.[103]

A few experiments have been done in Europe and America on displacing migratory birds and following their subsequent paths by recoveries of ringed individuals or by sight records. The best is that in which 144 young white storks (*Ciconia ciconia*) were transferred from Rossitten in east Prussia to Essen in West Germany.[314] Storks breed over much of northern and eastern Europe, but are divided into two populations, a western which migrates southwest and winters in northwest Africa, and an eastern that migrates southeast and winters in east Africa. The experimental birds would normally have flown southeast; they were detained until the Essen birds, which would have flown southwest, had left, and then released. More than 100 were subsequently collected or seen, along a line that was generally to the southeast, which took them into the unfavourable conditions of the Alps (Fig. 7.20). They must have been following an innate preferred direction.

Other experiments, on starlings, hooded crows, American crows (*Corvus brachyrhynchos*), white-collared flycatchers (*Muscicapa albicollis*), sparrowhawks (*Accipiter nisus*) and mallards gave similar but sometimes less clear-cut results.[310] In most experiments some birds continued on their preferred direction, but others were recovered within their normal range, so that they had apparently been able to reorientate. Since the figures have not been statistically analysed it is possible that there was merely chance return of birds that had failed to follow the standard direction. There is some indication that adults find their way back to their proper area better than young birds, but again no proper statistical analysis has been done.

About 20 years ago Gustav Kramer claimed that birds in a state of migratory restlessness showed their preferred direction in closed cages by facing predominantly that way, or flying backwards or forwards along a radius.[193] Other workers have said that they spend most of their time in the sector of the cage towards this direction, and it is not always clear what the observer in this sort of experiment has in fact recorded.

Accepting that the preferred direction can be determined in this way,

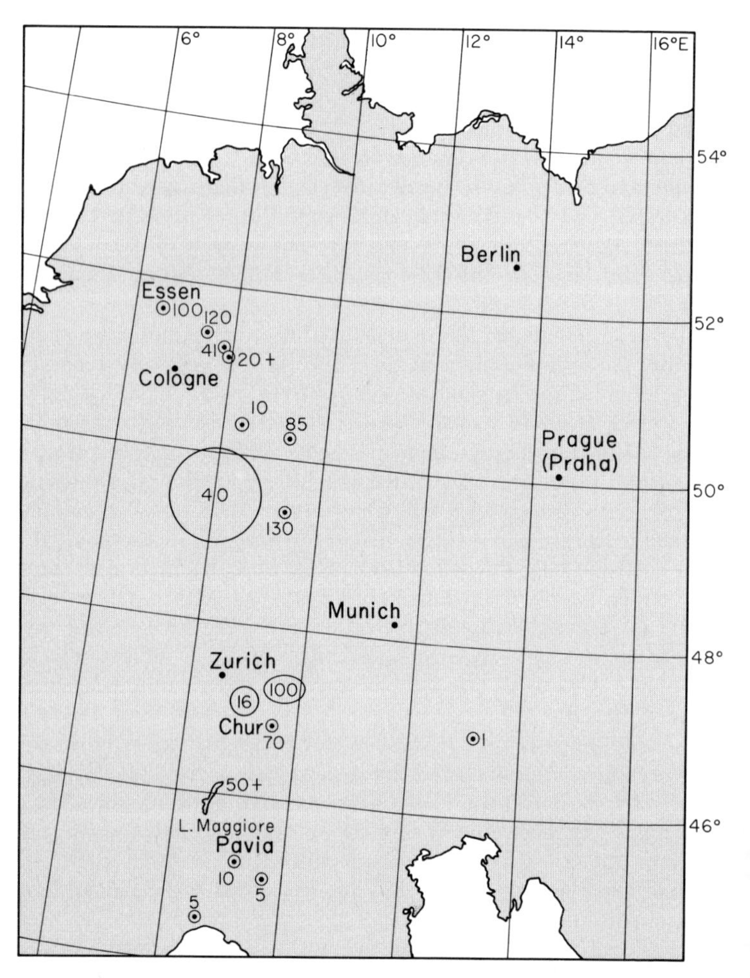

Fig. 7.20 Migration of storks. 144 birds released at Essen had been bred at Rossitten, 1000 km to the east. Figures show the numbers of birds recovered or seen. Scale: 10 mm=100 km. (Based on Schütz, 1958, *Proc. X int. orn. Congr.* 252.)

a number of workers have used the method to try to find out if the direction of fluttering of starlings, warblers and a few other birds can be altered by manipulation of the external conditions. Briefly, it has been claimed that it can be so altered either by using mirrors to give a false apparent direction to the sun, or by installing the cage in a planetarium and rotating the celestial dome. The conclusion is that the birds orientate by the sun during the daytime and by the stars at night, and this interpretation is supported by the loss of orientation when the cage is completely enclosed or under overcast skies. In one experiment however robins and whitethroats (*Sylvia communis*) retained their direction even when deprived of all visual clues, but lost them when placed in a steel chamber in a basement unless the door was open.[118]

The claimed results of these experiments have been generally accepted by ornithologists, but are nevertheless subject to grave doubt. Four types of criticism may be made. In the first place, often only one or two birds have given clear-cut results, and others that have not have been written off as uncooperative. Even in the best experiments many individuals have moved in the wrong direction, or not at all, which is very different from what happens under natural conditions.[100] Secondly, different observers have measured the orientation in different ways, but have got the same result. Thirdly, all the observations seem to have been made by people who knew the results they wanted; this is shown by the fact that the negative results have not, in general, been published. Finally, no measurements of other physical factors that might have affected the birds have been made. In the experiments in sunlight, for example, there must have been differences in the temperature of different sectors of the wall of the cage, and these might have affected the birds. For all these reasons the experiments cannot yet be used to support any theory of migration.

Migratory birds, like residents, often return to breed close to where they were hatched, and even more often to where they have bred in previous years, even sometimes to exactly the same nest-site. For the final stages of such a return, recognition of the area would seem the most likely method.

In summary, migrating birds seem to have an innate tendency to travel in a certain direction in spring, and in the opposite direction in autumn. These directions can be modified to some degree by obstructions and landmarks and are perhaps subject to learning. Forward and return journeys are not necessarily made over the same track, and the standard direction is not always the same throughout the migration. A high proportion of the birds achieve their goal, but many individuals lose their way and disappear. The means by which orientation is achieved remains unknown, but there is a suspicion (and a belief on the part of many ornithologists) that they migrate in some way by the sun or the stars.

It is possible, though not necessary, that the same system might be used in migrational navigation as in homing, on which experiments are much easier. They fall into three main groups, but in each the principle is that birds are carried to a point distant from their home and then released. They may, in exceptional cases, be followed by aircraft or telemetry; their time of arrival at the home may be recorded; or their direction of departure from the release point may be observed.

In one of the best experiments of the first type, gannets, which do not normally fly over land, were released in North America 90 miles from the coast, and their flight followed, as far as possible, in an aircraft.[138] Although only one of nine birds was actually followed to the coast, four returned to their home island within 75 hours; the paths of all of them, so far as they could be traced, were entirely random, except in so far as they were influenced by air currents. Presumably the birds struck the coast sooner or later by chance, and then found their way home through country that they knew.

In a longer series of observations on 10 homing pigeons fitted with radio transmitters 131 flights were made, in all but two of which the birds successfully returned home. The most striking thing was that although no bird deviated more than 10 miles away from the direct route of 35 miles, no two birds followed the same route. Some flights were nearly straight, others very jagged. The pigeons flew at tree-top height, so that there could have been no distant recognition of landmarks, and a bird sometimes crossed an area that it might have remembered from its training flights, without apparently recognizing it. Although they tended to sit when the sun was not shining, they homed well in haze, suggesting that a precise observation of the sun's position was unnecessary.[251]

In contrast, Adelie penguins (*Pygoscelis adeliae*) liberated on the featureless Antarctic icecap 341 km or more from their breeding grounds mostly moved on parallel lines, which, as close to the pole as this, were not all on the same bearing;[99] 103 birds were followed individually for 2–4 km. This type of movement would bring the birds back to the sea, and so would enable them to home by landmarks; a few of them did in fact return home. The standard direction was lost when the sun was not visible, suggesting that the birds were navigating by reference to its direction and the time of day.

Random exploration will explain even many of the spectacular returns of birds displaced great distances. Manx shearwaters (*Puffinus puffinus*) for example, have been taken from the coast of south Wales inland or to the Mediterranean, and a few have returned. It has been shown that their numbers are not higher than would be expected by chance, on a simple hypothesis of flight in straight lines alternating with turns, both the length of the line and the angle of turn being random.[395] One must assume in addition that the bird knows a moderate length of coast-line near its home.

The birds that navigate best are homing pigeons. They are usually trained over increasing distances, so that some knowledge of landmarks is possible, but they can also return in high proportions from places where they have never been before. Different strains have different abilities, and there can be little doubt that these are due to genetical variation. In general, the further the point of release is from the home loft, the lower the proportion of birds that return. The same is true of several species of wild birds that have been tested in this way; it would be expected on any theory of navigation.

Some people have suggested that birds navigate by the earth's magnetic field. This is theoretically just possible, though unlikely on present knowledge. Pigeons carrying magnets on various parts of their body, including their wings, where they would set up disturbing magnetic fields, home just as well as control birds, so the hypothesis may be taken as disproved.

Few birds when released fly steadily in one direction, but in several experiments their compass bearing from the point of release at the moment when they disappeared from view has been recorded, and taken as their initial direction. Some experiments of this sort show, or apparently show, that more birds disappear towards the direction of home than would be expected by chance;[240] for example for 340 pigeons in one series the average deviation from the home direction was 47°, the expectation if flight were random being 90°. Other experiments have not given such clear-cut results; in one, good orientation was shown at release distances less than 20 km from home or more than 100 km, but none from intermediate release points.[309] It is further claimed that the proportion of disappearances in the correct direction falls off when the sun is not shining. An obvious criticism of the experiments is that the atmosphere and the light are seldom uniform throughout the 360° of the compass, so that birds recorded as disappearing in different sectors will be at different distances. What effect this has will depend on how the bird has been flying before it achieves its definite direction, and on how long it takes for this. Different degrees of brightness of sky could themselves affect the direction in which the birds fly; indeed they would be expected to. In one experiment on Manx shearwaters that were released in Birmingham the majority undoubtedly departed in the direction of south Wales, which was their home. However, owing to the form of the city's smoke cloud the brightest part of the sky is also in that direction, so that the experiment proves nothing.

If one accepts, as most workers now do, that homing and migratory birds navigate by means of the sun, or, at night, of the stars, one must still ask what aspect of celestial geometry they use. There are various possibilities, and critical testing of them is not easy. The minimum that is required is a chronometer, and granting as we may that many animals do possess an internal clock, it would have to be of fantastic accuracy to account for some of the observed returns. According to one hypothesis, the bird, if it sees the sun moving, can determine the highest elevation that it will reach (which it will do at local noon) and the time at which it will be in this position by the internal clock. Comparison with the same values at the home (which must be remembered) will enable the bird to calculate the direction in which to fly to get home. Some experiments were claimed to show that it need see the sun for 10 seconds only, but in others pigeons that had been kept out of sight of the sun for eight days before release departed in the right direction as often, and homed just as well, as the controls.[164] A simpler hypothesis assumes that the bird knows only that the sun is higher or lower in the sky than it would be at home; if it is the former the bird flies towards the sun, if the latter away from it, and by continuous adjustments will gradually approach nearer home.[373] For this, and possibly for the other hypothesis, an artificial horizon would be necessary.

Neither of these methods of homing would seem to have much to do with migration, but if, by means of their chronometer, the birds could vary their angle to the sun, they could maintain a constant heading, just as did the penguins above. There is evidence that mallards, and some other birds, do set off predominantly in one direction (different for different populations) when released, whether by day or night.[241] Exposing them to artificial days moved their preferred direction in the way that would be predicted if their biological clocks had been reset so as to make them think the day began when the false day began in their cage, and not when it actually did. But at night their ability to move in the normal direction was unimpaired. A man could navigate in these circumstances by walking at a constant angle to the pole star.

8

Other Complex Behaviour

8.1 FOOD-SEEKING

At the beginning of Chapter 6 the feeding of chicks and tits has been used to illustrate the difference between innate and acquired habits. There can be no doubt that in nature the methods by which a species obtains its food are built up in the same sort of way, by instinct modified by learning, but without rigid experimental work, which is usually impossible, we can only guess at how far a particular habit depends on experience.

There can be little doubt that when woodpeckers tap the trunks of trees they receive information about the possible presence of large insect larvae beneath the surface, presumably through the sound that the tap produces. Blue tits and coal tits in pinewoods in winter feed heavily on the larvae of a small moth, *Ernarmonia* (*Laspeyresia*) *conicolana* (Heyl.), that lives in the scales of the cones.[128, 129] The tits' method of feeding is, to begin with, a general searching, which is presumably instinctive, but they seem to be able to find an occupied cone very quickly, since they examine each one for only about five seconds. The scales are sometimes tapped, but since there are usually only one or two larvae per cone and there is not enough time for the bird to tap all the scales, it seems unlikely that much information is got in this way. Occupied scales are rarely recognizable from outside by man. We do not know whether the recognition of parasitized cones is innate, or whether it has to be learnt afresh by each individual. If there is learning it must be relatively quick, and may be compared with that of a hand-reared grey squirrel (*Sciurus carolinensis*), which found nuts by smell, presumably instinctively, and at some time between the age of eight months and 34 months had learnt (probably by weight) to distinguish good nuts from those which were empty because the kernel had been destroyed by weevils.[212]

Bullfinches are well known to attack buds, but it seems that they do so only when not enough seeds, which are their usual food, are available, and trials of small numbers of birds in aviaries suggest that, in cold weather, they cannot maintain their weight on a diet of buds alone.[272] They make a distinct choice, not only preferring flower buds to leaf buds, but one variety of fruit to another; Conference pears are heavily attacked, while Doyenné du Comice, generally rated as the best of all dessert pears, are usually left alone.

Shelduck (*Tadorna tadorna*), at least on the shores of Kent and Somerset, feed almost entirely on the small snail *Hydrobia ulvae*, with small quantities of related species, although many other animals are available.[280] During the breeding season, however, when they may nest far from the sea, they must eat something else.

In all these cases recognition of suitable food by smell would be an obvious explanation, but is anatomically improbable, especially in the passerine birds.

Abnormal quantities of food attract large concentrations of birds; outbreaks of defoliating caterpillars draw various insectivorous passerines to the woods, and dead bodies cause the normally solitary vultures to collect in numbers. One of the most striking sights in tropical Africa is the assemblage of birds round bush fires. Bee-eaters come in numbers to catch the large insects driven on to the wing by the heat and fly almost into the flames to do so, while several species of raptor sit on surrounding trees waiting for small mammals. This behaviour is unlikely to be an instinct, and is presumably learnt afresh by each individual, no doubt aided by imitation.

Some birds have evolved peculiar methods for dealing with difficult food. The large beak of the hawfinch enables it to split the seeds of hawthorns neatly open, and it feeds commensally with thrushes that eat the fleshy fruits and drop the seeds on the ground. Crossbills feed almost exclusively on the seeds of various conifers, which they obtain by means of their odd-shaped beak. According to the best account the beak is opened, so that the tips come together, and is inserted between the scales of the cone. The beak is then partially shut, which presses the curved tips apart and so forces the scales away from each other; in this position the seed is revealed and picked out by the tongue.[409] As breeding birds, crossbills are tightly bound to coniferous forest, but they can if necessary eat other food, for example apple pips, which they can extract from the fruit. The red-throated bee-eater (*Merops bullocki*) squeezes out the venom of stinging species of insect (including drone bees) but eats other insects, including some that mimic bees, without this treatment.[119]

The oyster-catcher feeds on, amongst other things, the edible mussel (*Mytilus edulis*), a large mollusc which it cannot swallow whole. It has two methods of attack.[139] When the tide is out it may hammer the shell until it is broken and the flesh can be extracted, and if the mussel is under water and the valves are gaping, it may insert its bill and cut the adductor muscle that pulls the two halves of the shell together. The two methods are apparently used by different individuals, and their relative frequencies are different in different districts. This could be explained by genetic variation, or, more probably, by chicks learning by imitation of their parents.

8.11 Storage of food

Several species of bird store food when it is abundant. The crested tit (*Parus cristatus*), coal tit and willow tit all hide seeds, and to some extent insects, in autumn.[149] The methods and places used differ somewhat between the three species, the most elaborate being those of the crested tit. It hangs beneath a twig, sticks the food to the underside with saliva, and then covers it with lichen, which also stays in position. There is no evidence that the birds remember where they have stored food, but they search later in the winter in the sort of place where food may have been left, and analyses of stomach contents show that they succeed in finding some of it. In some willow tits in winter not less than half the food consisted of material that was not then available in the fresh state, and so must have been obtained from stores. The crested tit uses the whole tree, the willow tit the trunk and large branches, the coal tit twigs, so that each species benefits chiefly itself by the habit, but there must be some general advantage to the genus. The marsh tit and Siberian tit (*P. cinctus*) also store, but the great tit and blue tit rarely or never do. Those that store are, with the exception of the marsh tit, the species that go further north and inhabit the less rich woodlands, so that presumably they are more likely to suffer from shortage of food in the winter.

The nutcracker (*Nucifraga caryocatactes*) buries hazel nuts and pine seeds, and the jay (*Garrulus glandarius*) buries acorns.[133] The jay pulls the fruits off the trees, and carries one to three in its gullet and one in its bill into open ground up to half a mile from the woods. Each acorn is buried, hammered in, and covered. Unlike the tits, both jay and nutcracker seem to know where to look for their stores; jays begin to dig them up only a week after burial and may continue until the following summer, after the seeds have germinated.[39]

8.12 Tool-using

Several birds show feeding behaviour that in a general way may be called tool-using. Song-thrushes smash shells of large snails (*Helix* and *Cepaea*) on stones used as anvils. Great spotted woodpeckers and nuthatches (*Sitta*) carry hazel nuts and wedge them into crevices of rough bark, and there hammer them to get at the kernels. Several large birds (crows, gulls, vultures) drop hard prey ranging from molluscs to bones and tortoises and feed on the soft contents that emerge if the outer covering breaks, but there seems to be no selection of hard ground on which to drop. On the Galapagos the finch *Cactospiza* (*Camarhynchus*) *pallida* breaks off a spine from a plant and uses it to probe in bark.[252] If it thus disturbs an insect it drops the spine and catches its prey. A local population of brown-headed nuthatches (*Sitta pusilla*) in Louisiana has been described as using chips of bark to prise other pieces of bark loose.[260]

8.2 FLOCKING

One of the most characteristic features of many birds is their habit of living in aggregations of many individuals. These are broadly of two sorts: those that occur only or primarily in the breeding season, and those that are formed outside it. The former are generally called colonies, the latter flocks.

8.21 Colonial nesting

Colonial nesting is shown by many sea-birds, such as auks, petrels and gulls, but also by some land-birds, such as rooks. Sometimes at least the origin of the habit lies partly in the limited supply of the type of nesting-site that the bird's life demands. If, like guillemots and fulmars, a bird nests on ledges on nearly vertical rock, its distribution is obviously limited, and concentration of nesting would clearly favour survival. Since the whole sea is before them no question of feeding territory is likely to arise. Birds leaving their nesting-ledges to feed do so individually, and many of these sea-birds are solitary outside the breeding season, except when they collect together over a shoal of fish, as gannets do. The rook also, which nests in tall trees but feeds in open country (in nature, chiefly grasslands), is limited to what must, before agricultural clearance, have been a very limited biotope. Only the grass steppes, with wooded river valleys, are likely to have provided suitable habitat, and concentration into colonies in narrow belts of trees for breeding would have been advantageous. In other species, however, this argument cannot be used, and other more general advantages must be sought.

When cliff-nesting sea-birds are spreading their range, small numbers of birds often occupy new sites for a year or two without laying eggs, and it has been suggested that social stimulation, which can be given only by a minimum number of birds, is necessary for ovulation or spermatogenesis. Against this it may be said that the new sites are likely to be less suitable and to be occupied by less vigorous birds, and there is evidence that the kittiwakes (*Rissa tridactyla*) that form new colonies are young birds that would not breed successfully anyway.[70]

Black-headed gulls, herring-gulls and lesser black-backed gulls (*Larus fuscus*) sometimes nest regularly close together, in colonies that may contain thousands of birds, on various types of open ground that are not, so far as one can see, very scarce. The chief cause of death in the chicks is predation, largely by other gulls,[153] which may be away from the main colony less. Protection against predators seems in fact to be the most likely general advantage of colonial nesting, for while the colony must be more conspicuous than an isolated nest, the ratio of predators (which are more or less territorial) to nests must be lowered, and there is some possibility of communal attack on the approaching predator. In the cliff-nesting kittiwake, however, the lower nesting success of those on the edge of a colony did not appear to be due to predation.[69]

8.22 Winter flocks

Winter flocks are characteristically formed by the association of birds that nest separately, and thus often consist of several species. After the young have left the nest families of most species of tit stay together, and within a week or so have joined other families of the same species; in another week they have joined other species, and the flocks so formed may be fairly stable throughout the winter. Sometimes, however, they break up or reform; the long-tailed tits are particularly prone to leave the others. There are usually many individuals of a species of tit in a flock, but they are often joined by one or two (seldom more) nuthatches, tree-creepers (*Certhia familiaris*) or lesser spotted woodpeckers (*Dendrocopos minor*). Within the flocks there is usually a rough separation of the species by position; great tits move through the undergrowth and search the leaf litter, blue tits work at a higher level amongst the trees, and coal tits perch chiefly on the boles and larger branches.

Comparable mixed parties of birds are found in forests in Africa (where they may include European migrants such as warblers and spotted flycatchers),[397] in Asia[221] and in North and South America.[72]

Some of these flocks occupy a fairly limited tract of country, which is not, however, defended, and so is not conveniently called territory. Sometimes, especially in bad weather, the flock moves further afield. Within its usual area its movement is apparently random[406] (Fig. 8.1a) and is remarkably similar to the searches of herring drifters[131] (Fig. 8.1b). Birds that hold territory in winter, such as the marsh tit, may join up with the flocks as they pass through; the same thing has been observed in Rhodesia.

Flocks may reduce, like colonies, the chance of attack by predators, and in European birds there seems to be some interspecific recognition of

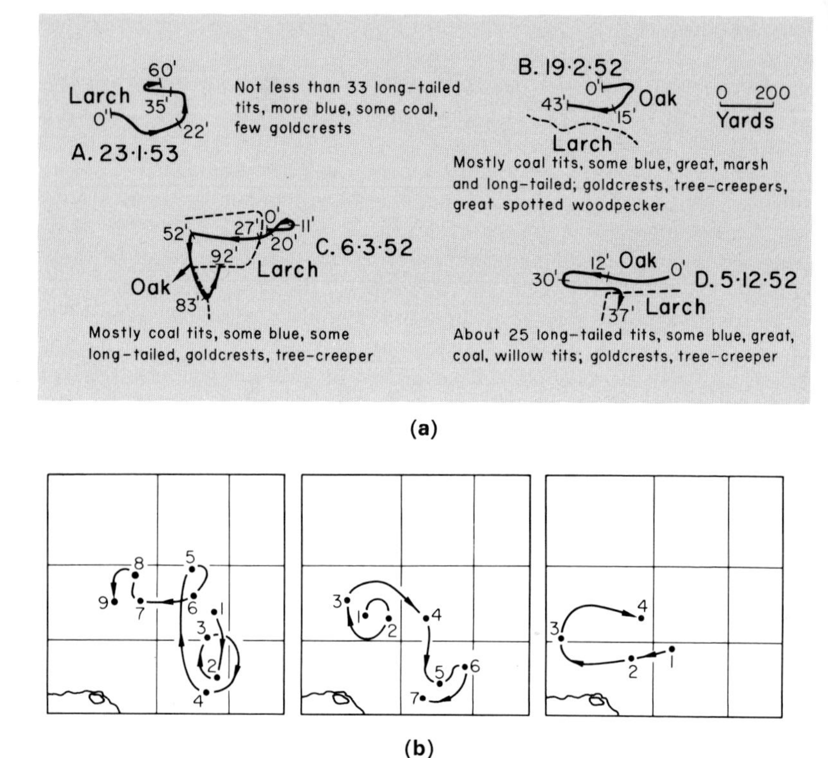

Fig. 8.1 (**a**) Tracks of tit-flocks. The dotted line shows the boundary of the type of woodland. (From W. B. Yapp, 1962, *Birds and Woods,* Fig. 13, Oxford University Press, London.) (**b**) Movement of a herring fleet. (Glover *et al.,*[131] 1956, Fig. 7, *Bull. mar. Ecol.* **4**, 152.)

alarm calls; in Rhodesia this was not observed. On the other hand flocks, by their continual calling, which seems necessary to hold the flock together, draw attention to themselves, and it would seem to be easy for a predator to take one bird from the edge of a flock. It has also been suggested that in Africa the primary value of the flock is that it flushes insects and so makes them easier to catch.[355] This can hardly be true of Europe, where the insects are not flying in the winter, and have to be searched for and picked off the branches of trees.

8.23 Roosting

A third type of association is for roosting. This perhaps begins as a search for warmth, since several wrens, which are solitary during the day in winter, may crowd into a single hole to sleep. Small feeding flocks of finches collect in larger groups for roosting, usually in thick evergreens such as conifers, holly or ivy, which presumably have a warm microclimate. The extreme of this sort of thing is shown by the starling, where small feeding flocks collect together in one large roost of thousands of birds, sometimes flying as much as 30 miles each evening to do so. Their invasion of a number of towns, chiefly with classical buildings, for this purpose suggests that the advantage is warmth. It can hardly be protection from predators, for sparrow-hawks have learnt that the assembling birds are easy prey.

8.24 Behaviour in flocks

Whatever the value of flocking, the behaviour of birds within the flock is an interesting subject that has been little studied. If a food supply is concentrated and limited, as when it is put out on a bird table, especially in bad weather, there is something of a peck-order, both within and between species, one bird being dominant to another.[46] But the order is not constant, and little of this sort of behaviour can be seen in the ordinary movement of a tit-flock through a wood, presumably because each food item is small enough to be swallowed as soon as it is found.

Flocks often move as one. Sometimes it can be seen that one or a few individuals of such species as curlews (*Numenius arquata*) take wing, and none follow, on which the fliers return; but if more than a few leave the ground the others all rise and the whole flock moves off. Who or what gives the signal for the wheeling of flocks of starlings or waders is unknown, but they move with great precision. The extreme of this is perhaps seen in small flocks of pelicans (*Pelecanus*) flying low over the water. Their wings beat exactly in step, and at irregular intervals they all stop as one and the birds glide, shortly afterwards starting to flap their wings again simultaneously.

It has been suggested that birds are basically gregarious, and that the separation by territory and individual distance is a later acquisition.[392] Since reptiles are solitary (or at most live in groups of one male and several females) any such gregariousness must have been early evolved because of the advantage it conferred.

8.3 MOBBING

Several species of bird, whether gregarious or solitary, have the habit of mobbing other larger birds, and temporary flocks are sometimes formed in this way. The bird mobbed may be a predator, when the habit may be useful in driving it away, but it is probably in England more often a species that seldom or never eats birds, such as a tawny owl (*Strix aluco*),

Fig. 8.2 Starlings sunbathing with tail fully spread and wings extended. Bare ground with cover close by is much favoured for sunbathing. (Photograph by C. W. Teager, courtesy of *Amateur Gardening*.)

a kestrel (*Falco tinnunculus*) or a cuckoo (*Cuculus canorus*). The owls seldom take any notice and are not driven away. One cock blackbird in a garden regularly mobbed a tawny owl every evening before he went to roost, even though the owl, roosting in a thick horse chestnut, had not been calling and must have been invisible. Such a useless habit seems inexplicable except in terms of human emotion such as annoyance. It may, indeed, be dangerous. Tawny owls feed almost entirely on mice and earthworms when they live in woods, but in London they live almost entirely on small birds; it is in gardens that mobbing is most intense.

8.4 SELF-STIMULATION

Several species of birds indulge in some form of self-stimulation. They may bathe, splashing water, snow or dust over their wings and feathers; they may sunbathe, spreading their wings or exposing their under-surface to the radiant heat[359] (Fig. 8.2), and more than 200 species of passerine pick up ants and apply them to the wing-tips (or sometimes to other parts of the body) or wallow amongst ants on the ground.[321] In all these activities the pattern of movement is more or less constant for the species, and so is presumably innate. Anting birds use only species of insect that produce formic acid (which seems to be the primary stimulus) but do not bite. Each individual has to learn which species to use, and it is perhaps this learning process that leads some individuals to use ant-substitutes such as cigarette ends and moth balls.

All these odd pieces of behaviour are usually said, for want of any better explanation, to be concerned in feather maintenance. Formic acid and dust may kill or dislodge ectoparasites, and sunbathing is said to convert oils into vitamin D (as it does in man). Birds normally keep their feathers in order by preening with their beak, which consists of re-zipping the hooks and hamuli on the barbules and spreading oil on them from the preen gland situated at the base of the tail. Bathing is said to help this oiling, but it seems doubtful whether feathers are ever really wetted, since if those of a duck are treated with a wetting agent the bird sinks at once. Dust-bathing is found chiefly in birds of open country, such as game-birds and larks, but these do not necessarily live far from water. Few species bathe in both water and dust; the house-sparrow and wrens do.

9

Distribution and Dispersion

9.1 STABILITY OF NUMBERS

Birds are, by any method of assessment, a successful group. There are many species, the number of individuals is high, they are conspicuous, and they exploit almost all types of habitat except the depths of the oceans and deep lakes. Their relationship to their environment must therefore be close and good.

One way in which this appears is in the stability of bird numbers. These can be estimated in various ways, but censuses in the strict meaning of the word can be made only for relatively large and conspicuous species, most so-called censuses being estimates based on some form of direct or indirect sampling. The world total of breeding pairs has been assessed with anything approaching accuracy only for the gannet, and for a few rare species. The gannet is an easy bird to count since it is large, and nests only in colonies in the north Atlantic, most of which are in Scotland and Ireland. In the early 1930s a count of most of the world's gannets, and an estimate of the others, suggested a total of about 80,000 breeding pairs, of which 54,000 were in the British Isles.[95] Near the other extreme, Kirtland's warbler (*Dendroica kirtlandii*), which breeds only in a restricted area of central Michigan, has about 1000 individuals,[242] and the whooping crane (*Grus americana*) about 30, some of them non-nesters, in Wood Buffalo National Park in the Canadian Northwest Territories.[225, 277]

If one cannot estimate the population in the whole world, the smaller unit that is most often taken is a country or some lesser administrative unit such as a county or part of one. Only for areas such as this, and again only for a few species, have we any estimates stretching over several years.

Most authors who have published such estimates have been concerned to point out any increases or decreases they have observed, and to explain them. In a longer view the interesting thing is how small the changes are. It is a commonplace that any continued excess of births over deaths, however small, will by itself lead to a geometrical increase in the population, and that any continued excess of deaths over births, however small, will by itself lead to extinction. A constant population must mean that birth rate and death rate are equal. If they are not equal in any one year, but equal over a longer period, say a decade, the population will fluctuate about a mean, and this is what the estimates for most species show.

Few estimates can be treated statistically, and even fewer have been so analysed, nor can statistics give one any information about systematic errors in one's measuring instrument. The justification for relying on most of the estimates of bird numbers is that they are self-consistent. They are certainly subject to error, though we do not know what it is. When that is taken into account the near-constancy of the recorded numbers of birds becomes the more remarkable.

The heron (*Ardea cinerea*) is a bird which makes large nests in conspicuous colonies, so that it is easy to count the nests, and so the breeding pairs. An attempt was made to do so over the whole of England and Wales in 1928 and again in 1954; the numbers were 3660 and 4281, the second figure being higher by 17 per cent. In intermediate and later years counts were made of selected heronries, and on the assumption that these were representative, estimates of the percentage change from year to year were made.[273, 339] Some results are shown in Table 9.1.

Table 9.1 Estimates of heron nests in England and Wales, recalculated from Nicholson (1938)[273] and Stafford (1969)[339].

Year	Number of nests
1928	4000
1942	2700
1945	4150
1954	4700
1955	4450
1956	4050
1957	4300
1958	3900
1960	4000
1961	4400
1962	3520
1963	2120

The rook, like the heron, nests in conspicuous colonies, and although there has never been complete coverage a number of small areas have been counted for several years running (Table 9.2). These and other counts suggest an increase over most of England in the mid-1940s, which is usually ascribed to the increase in arable land brought about by the war-time agricultural policy. The decrease shown in Cheshire since the war may be connected with the fact that this area, originally largely rural, has been increasingly suburbanized and industrialized.

Table 9.2 Number of rooks' nests in some English counties, 1928–68.

	Wirral 159	N. Chesh. 159	Derbys. 218	Notts. 85	Leics. 35	Herts. 307	W. Glos. 405	Corn-wall 67
1928				6576	9381			
1929	1704							
1932				6113				
1933							1009	
1934							1059	
1935							1075	
1936							1052	
1937							1126	
1938							938	
1939							1113	
1944	1647		16 114	10 306	13 639	8432	1204	
1945	1365	1595						1578
1946	1382							
1947	1197	1381						
1949	1124	1028						
1950	1238	1300						
1951	1329	1305						
1952	1434	1459						
1953	1360	1289						955
1954	1342	1165						
1955	1266							
1958				17 028				
1960	860					*c.* 16 000		
1961	748							
1962	806			10 609				
1963	870							
1964		617						
1965	787	507			10 652			
1966			12 630					604
1967	638							
1968	648							

The great crested grebe was counted over almost the whole of England and Wales in 1931 and 1965, and enough sample waters were counted from 1946–55 for an estimate of the total population to be made[291] (Fig. 9.1). If the recent rise is more than a fluctuation it is probably due to an increased number of suitable breeding waters, mainly in flooded gravel pits.

In a different way, some attempts have been made to estimate the total population of all the birds in an area. The most ambitious of these is the series of Christmas Day bird counts organized by the National Audubon Society of America.[239] It differs from the others in being made in winter, when day to day fluctuations would be expected to be large because of the movements of migrants caused by changes in the weather. The mean number of birds recorded by 5160 observers for 1949–52 was $8{\cdot}6 \times 10^6$, while in 1962 9981 observers recorded $44{\cdot}6 \times 10^6$ birds.

The reporter thinks that this suggests a real threefold increase, but this may be doubted. Sample counts in the breeding season in Illinois in 1909 suggested that there were about 61,000,000 birds in the State, while similar counts in 1957 gave a total of 60,000,000, a remarkably close agreement when one considers the changes that have taken place in the use of land.[134]

The London Natural History Society has counted singing males in 16 hectares (408 acres) of oakwood in Surrey over 15 years[33] (Fig. 9.2). In view of the small numbers, the totals must be considered remarkably constant, except for the willow-warbler; what fluctuations there are can to a great extent be correlated with hard winters preceding the counts. Line transects made over roughly the same period in managed oakwoods of 152 hectares (380 acres) in another part of England showed greater fluctuations, some of which could be ascribed to the forestry operations that the woodland had undergone. Nevertheless, for some species there was again reasonable constancy[408] (Fig. 9.3). Counts in three different types of wood in Finland from 1948–52 showed similar rough constancy.[274a]

There are some species that have undergone great reductions or increases, which generally become obvious in any country with organized natural history societies. Some such changes have been fairly fully documented; sometimes a reasonable explanation can be given, more often it cannot. Changes of this sort are most obvious and most frequent when a bird is on the edge of its range, and it seems likely that under such conditions the ecological requirements for the species, whatever they may be, are only just met, so that a small change in the environmental factors may have a disproportionate effect. On the other hand, a genetic change in the bird may enable it to exploit a wider environment and so extend its range.

9.11 Some decreases

The corncrake (*Crex crex*) has declined greatly during this century both in Finland and in Great Britain, so much so that over England and Wales

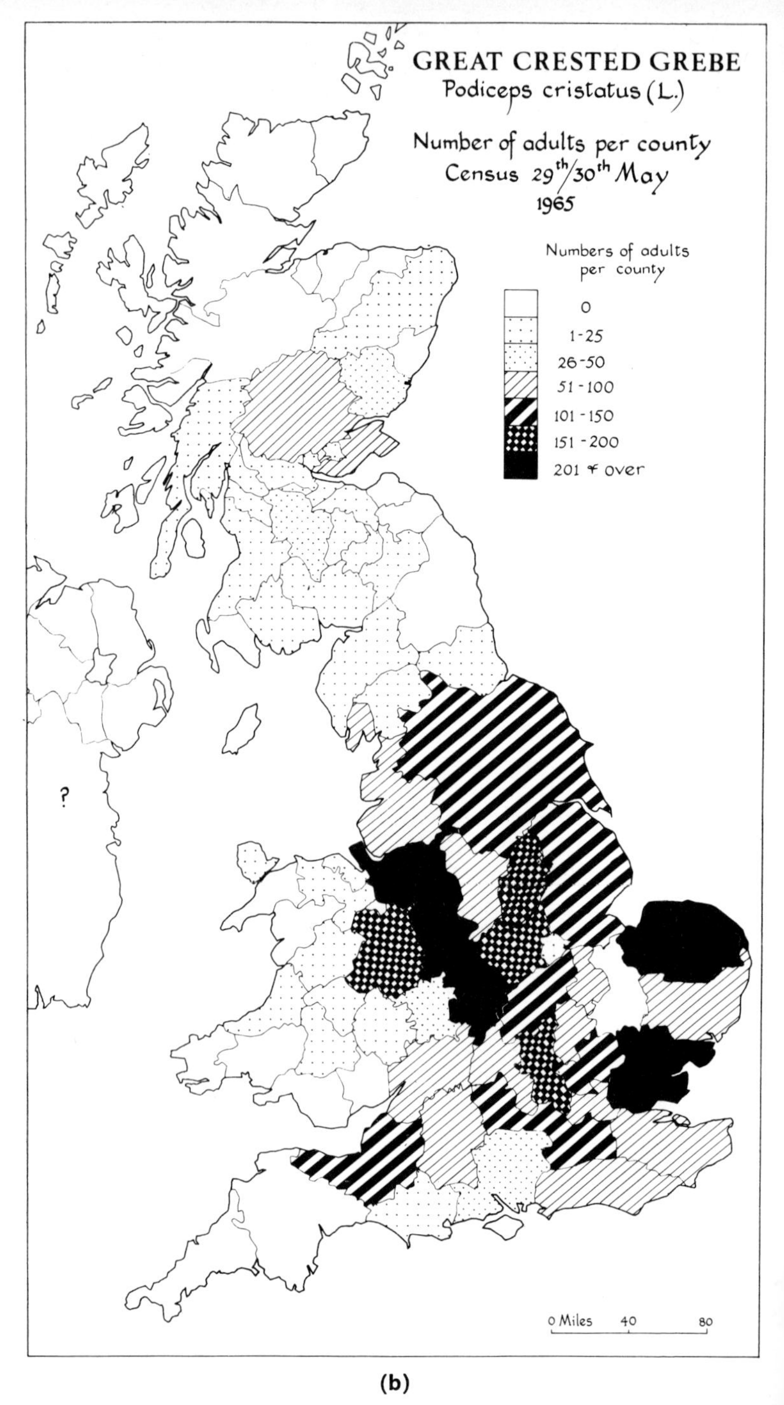

(b)

Fig. 9.1 Maps showing the numbers of great crested grebes in Britain (**a**) in 1931 (**b**) in 1965. (From Presst and Mills,[291] 1965, Figures 1 and 2, *Bird Study* **13**, 196–7.)

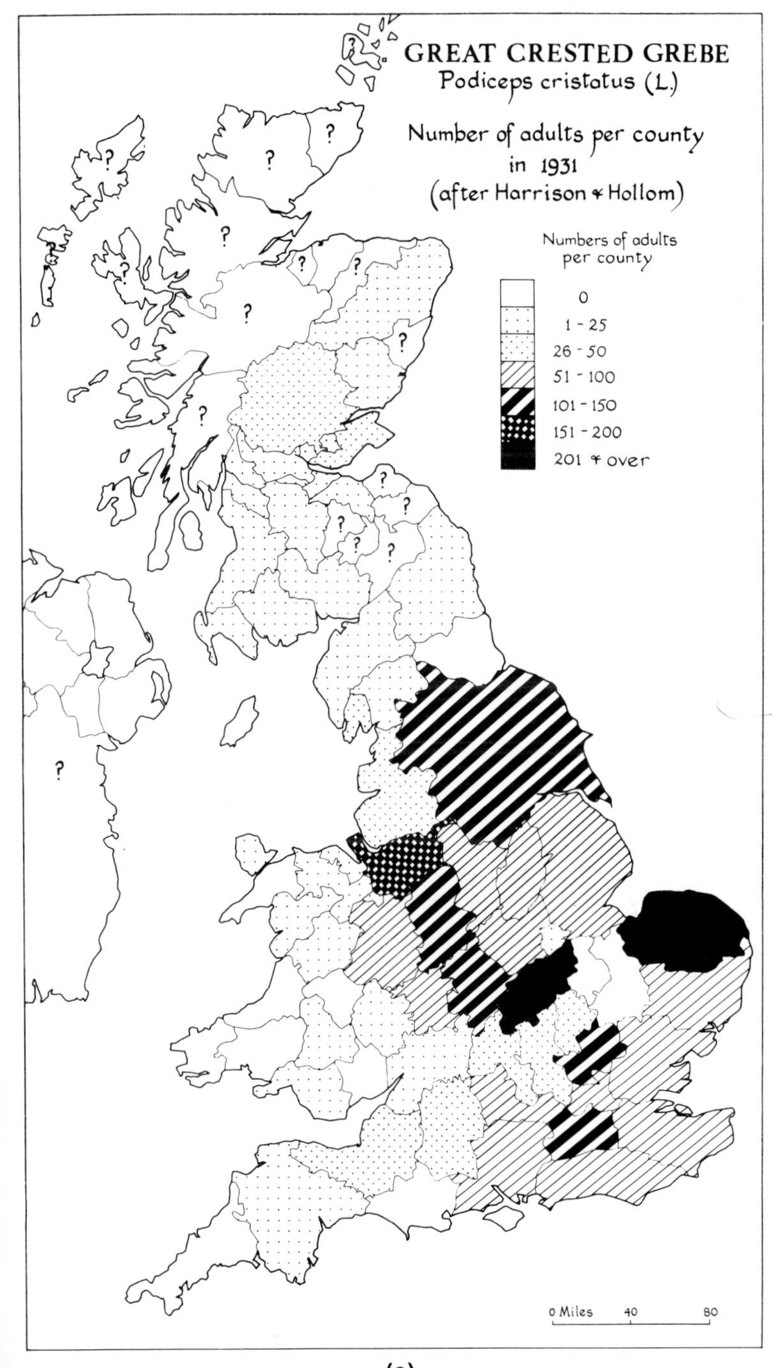

GREAT CRESTED GREBE
Podiceps cristatus (L.)
Number of adults per county in 1931
(after Harrison & Hollom)
Numbers of adults per county
0
1 - 25
26 - 50
51 - 100
101 - 150
151 - 200
201 & over
0 Miles
40
80

(a)

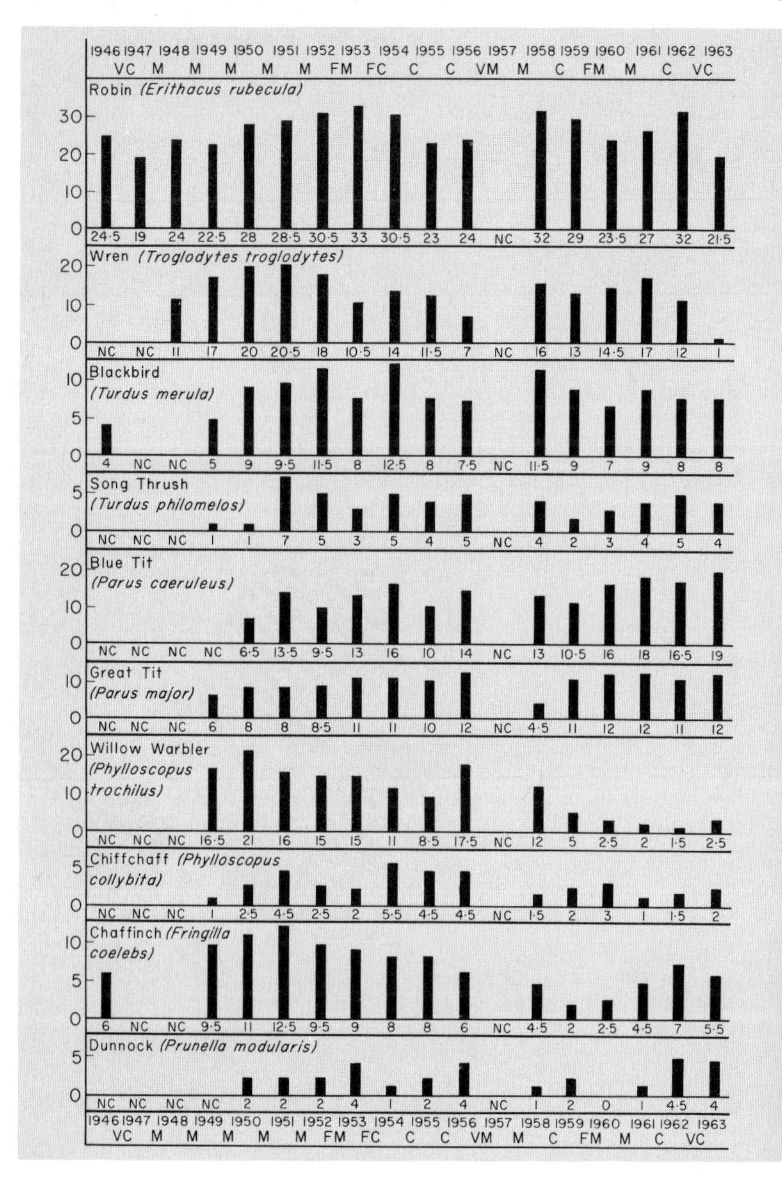

Fig. 9.2 The numbers of territories of singing males in Eastern Wood, Bookham Common, Surrey, from 1946–63. Numbers for each year are given at the foot: VC very cold; C cold; FC fairly cold; FM fairly mild; M mild; VM very mild. (From Beven,[33] 1963, Fig. 1, *Br. Birds* **56**, 307.)

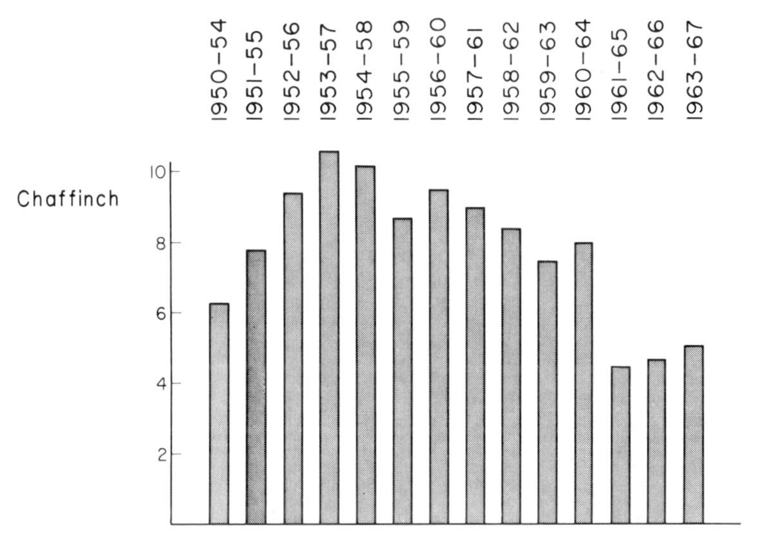

Fig. 9.3 Number of contacts made with chaffinches per hour in Wyre Forest, in summer. Two types of woods taken together (oak and beech, birch and oak). (From Yapp,[407] 1969, Fig. 9, *Proc. Bgham nat. Hist. Soc.* **21** (3), 199–216.)

it is, to all intents and purposes, extinct, though it could formerly be heard on farmland everywhere. The reduction can hardly be due, in Great Britain at least, to climatic changes, for though the country is near the northern limit of the bird's distribution the decrease began in the south and has on the whole moved northwards, so that by 1939 the bird was still fairly common in Scotland though rare in England. Fifteen years later it was, even in Scotland, almost confined to the extreme northwest and to the islands. Nor is the usual explanation, that its habitat has been destroyed by mechanical mowing and reaping and earlier harvests, convincing, for there are plenty of uncultivated areas in England similar to those in which it lingered in Scotland.

Some other birds that have declined in England have retreated towards the southeast, so that climatic change may be the cause. There is, however, no certainty of this, for though the past 20 years have seen cooler summers on the whole, the general trend over the past 150 years has been an amelioration of climate. But at least these birds have been retreating towards their main populations on the continent, which is what one would expect. Examples are the red-backed shrike, formerly widespread but now confined to the south and east, and the wryneck (*Jynx torquilla*), of which fewer than 30 pairs bred in 1966, mostly in the extreme southeast.[284]

The wryneck is said to have declined in many countries of Europe, but the red-backed shrike is still common on the continent. Both species are migrants and seem to prefer a continental climate.

The decline of the stonechat (*Saxicola torquata*) has not been so marked, but from breeding in every county of Great Britain at the beginning of the century it had fallen to the state where it was almost confined to the maritime counties of the south and west and to Surrey.[226] It is a resident bird, and this distribution suggests that it has been affected by hard winters, since the counties where it remains are those with the most Atlantic climate. There is some collateral support for this view.

There are some species whose decline can be confidently ascribed to human interference. They are mainly the birds of prey which in the nineteenth century were shot because of their supposed adverse effect on the numbers of game-birds. Once a bird became rare (and large predators are never in high numbers) egg-collectors who wanted rarities added a further destructive pressure. The osprey (*Pandion haliaetus*) was exterminated in Scotland in this way, and all the other larger falconiform birds were confined to the mountains on the coast.

Doubts have been cast on the commonly-believed decline in the peregrine falcon during the nineteenth century but there is good evidence of the bird's massive destruction, which is unlikely to have been entirely ineffective. However that may be, the evidence suggests that during the decade 1930–39, when the bird was protected, the population in Great Britain was fairly constant at about 600–650 pairs. During the war the bird was almost exterminated in England, because of its attacks on carrier pigeons, but not so much reduced in Wales and Scotland. By 1955 numbers were said to be back to normal in most districts, but in 1962, when more than half the known eyries were examined, no birds were seen at more than half of these (35 per cent of the total) and only 125 pairs were known to have laid eggs.[296, 297] This decline has been commonly ascribed to organochlorine pesticides, and the finding of these in the eggs is taken to be confirmation of this. But there are indications that there has been a decline in the Scottish Highlands that is independent of or antedates the use of such pesticides, and the case must be left as unproven.

The history of the buzzard (*Buteo buteo*) is different. The bird's requirements are less exacting than those of the peregrine: trees in which to nest, small mammals of almost any sort on which to feed, and open country in which to find them. It ought, therefore, to be well suited to the farmland of England, with hedgerow trees and small woods, and fields with plenty of mice, rats and rabbits. Historical records suggest that in 1800 it was ubiquitous, but that by 1900 its numbers had fallen very low, owing to destruction by gamekeepers.[254] Over lowland and agricultural England and Wales, and over much of Scotland, it was exterminated, but it persisted in fair numbers in the mountains and hills of the west and the north. Perhaps because the numbers of gamekeepers were reduced during the first World War it began to spread into the lowlands again in the 1920s, and by 1954 it was nesting regularly in some lowland counties and less frequently in others (Fig. 9.4). Then in 1954 the numbers crashed, and it disappeared once more except from the hills. This was the year in which the rabbit was almost exterminated by myxomatosis, and it may be that it was this loss of food that caused the drop in numbers. It is difficult to see why this should have been so, since there are plenty of other small rodents, and some surveys have shown that the buzzard's chief food is in fact the short-tailed field vole (*Microtus agrestis*). The buzzard was (and now is once again) certainly common on hillsides with hardly any rabbits. It happened that May 1954 was extremely cold, and as the buzzard nests in that month there might well have been many desertions. It is now once more present in some lowland districts, though not as common as it was in the early 1950s.

9.12 Some increases

Some species have made spectacular increases in population, but nearly always they have been accompanied by an increase in range. One of the biggest increases, although little known because the bird is taken for granted in most parts of the country, is that of the starling. It spread into the western counties of England and Wales and into most of Scotland only in the last century, and since migrant flocks come in the winter from Eastern Europe, the whole British population probably represents a gradual failure to return on the part of some of these visitors. Its spread in North America since it was introduced in 1890 has been even more rapid.

The fulmar breeds on cliffs, and, because it has been regularly taken for food, estimates can be made of its numbers for several years back. At one time its only breeding colonies in the British Isles were on St. Kilda, and it seems likely that from 1829 to 1939 the numbers there were roughly constant at about 20,000 to 25,000 pairs. Then there was a rise to about 40,000 pairs, at which figure the population remained. Meanwhile other cliffs on other islands and on the mainland of Britain and Ireland were colonized; the birds gradually spread south, and by the early 1940s had reached the English Channel from both east and west.[112] The total estimated British and Irish breeding population at 10-year intervals is shown in Table 9.3. Since there has been no diminution in the number of fulmars elsewhere, there has been a big increase in their total numbers. Two explanations have been put forward; that there has been an increase in the available food supply in the form of offal from whalers and trawlers, and that a genetic change that has raised the birth rate consistently above

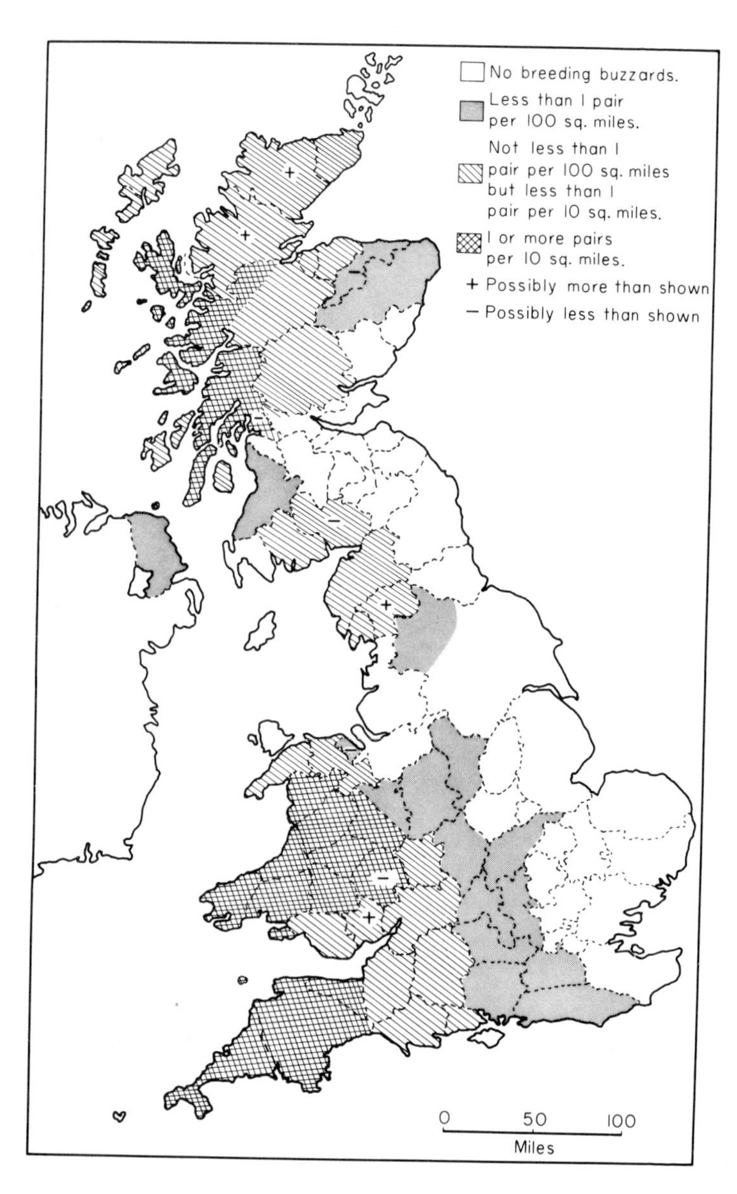

Fig. 9.4 Breeding density of the buzzard in Britain in 1954. Scale in miles. (From Moore,[254] 1957, Fig. 4, *Br. Birds* **50**, 188.)

Table 9.3 Estimates of breeding pairs of Fulmars.[112]

Year	St. Kilda	Rest of Great Britain
1829	25 000	
1879		24
1889	(probably	100
1899	stable)	480
1909		2000
1919		6000
1929	25 000	13 000
1939	21 000	35 000
1949	38 000	71 000
1959	38 000	97 000

the death rate has produced more wandering young that explore and inhabit new cliffs. The method of expansion is certainly that young non-breeding birds first explore a cliff, and then after a year or two lay eggs, but this does not tell us whether the birth rate has increased or whether, before the big supplies of fish offal became available, there was not enough food for all the young that were hatched. More will be said about this type of problem below.

The gannet, another cliff-nesting sea-bird, has also increased in numbers.[27, 113] The colony on Grassholm, off the Pembrokeshire coast, increased from 20 pairs in 1883 to about 5000 in 1937 and to about 15,500 in 1964, and that on the Bass Rock in the Firth of Forth from about 4800 in 1949 to about 7000 in 1962. The gannet does not normally feed on fish refuse, but takes only whole fish, so that the explanation given for the fulmar cannot apply. There was probably a decline in numbers of the gannet during the nineteenth century; the changes have been related to changes in persecution by man.

9.2 CLUTCH SIZE

The crude birth rate in birds may be measured by the number of eggs in the clutch, the term used for a group of eggs produced and incubated more or less together by one pair. In some species allowance must be made for second clutches in the same season, in others for irregular breeding, and adjustments must be made where there is polyandry or polygyny.

The number of eggs in a clutch is approximately constant for a given species (that is, it varies, not very widely, about a mode), and ranges from double figures in some small passerines and game-birds down to one in many sea-birds. Closely-related birds usually lay about the same number of eggs. Several general rules about clutch-size have been published, but all have exceptions. Most large birds lay fewer eggs than small ones, but the ostrich lays the relatively high number of five to ten. Hole-nesting birds lay more eggs than those that make nests in trees, but owls, which nest in holes, have small clutches. The same species often lays more eggs the further north it lives, but the rule is not as invariable as is sometimes claimed.[148] In years of good food supply clutch-size may go up; this applies very strongly to short-eared owls, which lay more eggs in years of vole-plagues, and to domestic birds, where high feeding means good egg-production (though there has been genetic selection as well). If these approximate rules mean anything, it is that the number of eggs laid is controlled by a variety of factors, and that attempts to reduce everything to one simple axiom are likely to be unsuccessful.

In spite of this, much argument and many assertions have been devoted to establish that the number of eggs a bird lays is determined by one single factor. With a stable population birth rate and death rate must in some way be mutually adjusted. A simple view of how this may be brought about is that the birds produce the maximum number of eggs that they can, and that the death rate rises to match this. If this were so, one would expect all birds to have large clutches and second broods, which they do not. A modified view is that they produce the maximum number of young that they can feed. This means that if they lay fewer eggs, there will be a surplus of unused food, and if more, some or all of the chicks will starve. This hypothesis can be tested by finding out what proportion of young survive to independence from broods of different sizes; the prediction is that it should be a maximum for the modal size of clutch, although this is not necessarily true with a wide spread of clutch size such as seldom occurs.[262] In practice, survival to independence is difficult or impossible to measure, so survival to the stage of leaving the nest is taken as the next best thing. In some species, such as the common and Alpine swifts (*Apus apus* and *A. melba*) the hypothesis was confirmed (Table 9.4). In some other species, such as the starling (Table 9.5), blackbird, song thrush, robin and redstart, it was not.[195, 196] These results can be explained away by assuming that mortality for the modal brood is less after the birds have left the nest, but explaining away results does not verify an hypothesis. Attempts to test the assumption by trapping a known population during their first winter have been indeterminate. At an ambient temperature of 12°C the oxygen consumption of nestling great tits was less in large broods than in small, so that, though they may get less food, they need less for survival.[249] It appears that the major part of the variation in clutch-size in this species is not genetically determined,[189] and it seems probable that litter-size in general is of low heritability.[105]

able 9.4 Survival in relation to brood-size in swifts.[196]

Brood	Number of young hatched	Per cent flying	Young raised per brood
	Alpine swift (*Apus melba*) in Switzerland		
1	58	97	1·0
2	562	87	1·8
3	1623	79	2·4
4	20	60	2·4
	Common swift (*Apus apus*) in England		
1	36	83	0·8
2	204	84	1·7
3	96	58	1·7

Notes: (i) In the common swift there was also one brood of 4, from which young flew; (ii) In both species the few late broods, which tend to be smaller, vere omitted from the analysis.

able 9.5 Nesting success in Dutch starlings (*Sturnus vulgaris*).[196]

Clutch	Eggs laid	Per cent flying	Average number raised per brood
3	105	82	2·5
4	864	84	3·4
5	2770	84	4·2
6	2184	82	4·9
7	756	82	5·7
8	112	78	6·2

Note: Losses of complete broods, due mainly to predation, were excluded, but they are included, the percentage of young flying is still constant with respect to rood-size.

Another line of approach is to associate the time of breeding with the ood supply. It is obvious that there must be enough food for the young grow at the time when they are produced, but the argument goes eyond this and claims that, on the average, the young are in the nest when ost food is available, so enabling the maximum number to survive.[75, 211] hat this is generally true was realized by Darwin.[76] Confirmation of it as been claimed for the blue and great tits, whose peak time for nesting aries from year to year and agrees with peak production of caterpillars n which they feed their young;[126] for the stock-dove, which feeds on the eeds of weeds;[267] and for the wood-pigeon, whose peak period coincides ith the August production of agricultural cereals, on the seeds of which,

in the district where the bird was studied, it largely feeds.[267] This las
case implies very rapid evolution of the peak nesting time, for cereals
have only been grown in Great Britain for a few thousand years, and can
only be the chief food of wood-pigeons, which are woodland birds, in the
limited parts of the country where grain is an important crop and woodlands are few. The deduction that the abundance of food determines the
breeding time is in any case illegitimate; an equally possible hypothesis is
that birds feed their young on the food that is most readily available.

Against the hypothesis there is much general evidence.[142, 324] If i
were true, species where only one sex feeds the young should have smalle
broods than related species where both sexes feed them, but there are
many examples where this is not so. If insectivorous birds such as tits
can raise large broods, as they do, there seems no reason why other success
ful birds, such as the chaffinch, should be restricted to broods half th
size. It is difficult to believe that sea-birds are really restricted by food t
rearing a single egg. Fire-finches parasitized by weavers in Africa, an
song-sparrows parasitized by cowbirds in America, regularly raise th
double brood, of their own eggs and the other bird's;[257] their norma
clutch cannot therefore be the largest they can raise, unless in fact the
have adapted to the parasitism by laying fewer eggs.

A few direct experiments have been done by adding eggs to a clutc
and then following its success. Those on great tits were inconclusive, bu
those on gannets have given better results.[271] The bird rarely lays mor
than one egg, and a second egg was therefore added. The proportion c
these twin-eggs that hatched was normal, and although the proportion c
the chicks that left the nest was slightly reduced, the number that did s
was increased by 76 per cent over what would have been expected fro
the same number of parents with single eggs. Clearly the bird coul
raise more young than it normally does. As has been said above, the ganne
is increasing in numbers even with broods of one. If it laid two one migh
expect the increase to be spectacular.

An alternative hypothesis is that, mortality being more or less fixe
the number of eggs laid and young produced is adjusted, by social facto
or group selection, to correspond to it.[94] There is a mass of general evidenc
not confined to birds, which suggests that this is a probable explanatio
but it is difficult to see how it works, and it seems to be incapable c
experimental testing. Many birds that lay only one egg, such as albatrosse
differ from the majority in that they do not, like most species, lay anothe
if the first is removed. Since their ability to rear young cannot be affecte
by such tricks that are played on them by man or other predators, the los
of this ability cannot be explained as an adaptation to the food suppl
but it might be part of the general social adaptation of low fecundit
necessary in a long-lived bird with low mortality.

).3 MORTALITY

If the birth rate is not adjusted to the death rate, the reverse may be rue. The more individuals there are the greater may be the proportion hat die. If this is so, mortality is said to be density-dependent.

It is easy to think of examples where the factors of mortality might ary in this way. If there is enough food only for x individuals, and y are resent, the proportion that die, $(y-x)/y$, will increase with y. That is the nathematical assumption, and there is evidence that in some natural ituations it applies. Birds feeding young in the nest tend to give the food hey carry to the nestlings that reach up furthest, which will be those that re biggest, who have already had the most food. Additional young will ot get their share, and so the same number, and a smaller proportion, vill survive in a larger brood than in a smaller. The same may apply to dults feeding on a limited but concentrated food supply, where peck-ominance will prevent the subordinate birds from getting their fair share. t will not necessarily apply where food is widely distributed in small uantities. If the number of tits in winter flocks in woods were increased, is certain that all would get some food, and they might get enough, but is possible to imagine a situation where none would get enough because, s a result of more birds eating the insects, the latter became so thinly istributed that finding them took too long. In the first stage mortality vould not be affected by density, and in the second it would rapidly rise 100 per cent, after which no further rise in density could have any effect.

In the simple case, predation is inversely density-dependent, for with a xed number of predators a rise in the density of the prey leads to a reduc-on in mortality. If, however, as often happens, the rise in the prey leads a rise in the predators, the mathematical prediction is for an oscillation numbers of both. The same may apply to parasitism and death by isease. Some birds, especially game-birds, do undergo cyclic or oscillatory hanges in numbers, and may be controlled in this way.

Birds are, of all animals, the least likely to be amenable to simple odels. Not only does much of their behaviour, such as territory holding, ean that simple correlations of density cannot be made, but their mobility eans that they can often escape from changing conditions.

.31 Death from starvation

Most small birds raise three or four or more young to the stage where ey leave the nest and can forage for themselves, so that there are five six individuals which must be reduced to two by the next breeding ason if numbers are not to rise. For several species there is evidence of a gh mortality from what may be called accidental causes until the birds the year have achieved full size in their first autumn, and after that mortality levels out. It is still high, and evidence from ringing returns and other sources for several species shows that the expectation of life is only a year or two, and is statistically independent of age. In contrast to this, which appears to be the general pattern, the black-capped chickadee was found to have a low mortality from fledging to the time when the flocks broke up and territorial behaviour began. At this point there was a sharp drop in numbers, but whether from death or emigration could not be determined.[329]

There are a few cases, in addition to the swifts mentioned above, where mortality of chicks in the nest seems to be largely by starvation and is probably density-dependent. On the grounds that the juvenile great tits caught in the autumn were predominantly those known to have been heavier in the nest, it has been claimed that starvation is the chief cause of the mortality of the young after they have flown, and is the controlling factor in determining the population of the next summer.[197] A subsidiary factor is the crop of beech mast, which tends to be high every two or more years; survival is then good, but it is lower when the mast crop is poor. The usefulness of these results is reduced by the facts that the observations were made in a heavily boxed area, with a higher density than normal, and that although great tits are common in beechwoods, they live also (and in Great Britain predominantly) elsewhere.

In the Netherlands it was found that in cold periods great tits in pinewoods lost weight, while those in deciduous woods ate more and kept up their weight.[23] The prediction made above was therefore fulfilled. One might extrapolate these observations and say that a slightly greater density would soon lead to extinction of the population in pinewoods, but would take longer in deciduous woods, which are the preferred habitat of the species.

Flocks of wood-pigeons feeding on the ground also compensated for the lower density of food by walking faster, and so eating the same amount in a given time, but some birds could only keep up with the flock by not eating, and so might be expected to die of starvation.[268]

The numbers of the tawny owl depend much more closely on their food.[333] The mice on which they feed vary greatly in density; when it is low the owls suffer from a state of semi-starvation in which they may not breed, and if they do so there is high loss of both eggs and chicks.

Predators in general seem to be able to switch their food according to what is available. Owls do this to a great extent, taking the small rodents that are easiest to catch at any given time and place. Tawny owls in London have changed their diet almost entirely to birds.[34] The golden eagle also takes anything of suitable size, mammal or bird, and where not enough of these is available, as in the western Highlands, eats carrion.[4] Although it fed largely on rabbits where they were available, their removal by myxomatosis had no effect on its numbers or on its breeding success

Where a predator has only one type of food it is vulnerable to changes in the numbers of the latter. The Everglade kite (*Rostrhamus sociabilis*) feeds exclusively on a single species of large aquatic snail (*Ampullaria* (*Pomacea*) *paludis*), and since drainage has reduced the numbers of these in Florida the kite has become almost extinct there, although it persists in South America.[347] In the opposite direction the introduction of the mussel *Dreissena polymorpha* into the Lake of Geneva has been followed by a rise in the numbers of several species of duck.[125]

9.32 Death from predation

Although much attention has been given, in economic biology, to the question whether predators can control species that are harmful to man's crops, much less has been given to trying to find out whether they are important in this way in a state of nature. It has been shown that large broods of the great tit are more often taken by predators (grey squirrels, weasels, and great spotted woodpeckers, of which only the last two are natural constituents of the English fauna) than are small ones, presumably because they make more noise and so call attention to themselves.[287] Destruction of the brood is complete, and often the mother is taken as well, so that there may be a strong selection against large broods. Since the predator has to search for the nests, and since it has many other sources of food, this is a case where predation may be density-dependent, for the more nests there are the more are likely to be found.

Many birds' eggs and young in the nest are taken by predators, especially in woodland by the Corvidae. The frequency with which, in spring, birds' eggs may be found on the ground points to the common occurrence of predation, and it has been shown that in the wood-pigeon 97 per cent of the loss of eggs is in this way, and the total loss is 60–70 per cent of the eggs.[266]

There is evidence that a wide range of predators, both mammals and birds, when they are feeding on adults, take chiefly the old or the diseased; in other words, those that will shortly die anyway. From this it is argued that predation has no effect in controlling numbers, and there is good evidence that in some cases and under some circumstances this is so. The red grouse that are killed by foxes are chiefly those without territories, that would not be able to breed.[172] There is still the possibility of some control here, for if any of the birds holding territories were removed, their place might be taken by some of those without. However, so long as any surplus birds are left, predation would have no effect on next season's breeding population.

One might expect that a bird such as a sparrow-hawk or a merlin (*Falco aesalon*), which flies rapidly and takes its prey by surprise, would not be so restricted to the old or feeble. There is however evidence that the sparrow-hawk takes more young birds than old, presumably because they are less skilled at escaping.[37]

There can be no doubt that man has occasionally exterminated a species, as he did the passenger pigeon (*Ectopistes migratorius*) in North America and the dodo of Mauritius, but even he sometimes has less effect than he expects. More than 60,000 Laysan albatrosses (*Diomedea immutabilis*) were killed on Midway Atoll, and many eggs and chicks destroyed, between 1955 and 1964 in the supposed interests of the United States Air Force, but there were as many collisions between birds and aircraft as ever, and the maximum reduction in numbers of the albatrosses estimated from aerial surveys was 50 per cent.[111]

Man's introductions of predators on islands have occasionally led to extermination, especially of flightless birds, but more often the population stabilizes at a new lower density; if this happens, the predation is controlling.

That disease, in spite of the mortality that it causes, is not always on balance inimical, is shown by the well-known case of sickle-cell anaemia, a genetically-determined condition in man, in which the heterozygotes are protected against malaria in Africa. Parasitism also is not necessarily as harmful as might be expected. Nestlings of various icterids in Panama suffer heavy mortality from botflies (*Philornis*), but they are protected from these when a giant cowbird (*Scaphidura orizivora*) lays eggs in the nest. The young cowbird preens its host sibs (which it does not eject, as the cuckoo does) and removes the flies.[328]

9.4 LIMITATION OF NUMBERS BY SPACE OCCUPANCY

The importance of territory, or of a particular type of nest-site, has been discussed in section 7.31. Clearly if the size of territory cannot be reduced below a certain minimum, or if there is a limited number of nest-sites, such as holes for tits and pied flycatchers, either may limit the population to a maximum. Below this other factors might be limiting, or the the population might expand geometrically in accordance with the difference between birth rate and death rate.

Once all the suitable territories or nest-sites are occupied members of the surplus population do not necessarily die, but will find it difficult to breed. In most cases, however, whether they breed or not, they will have to occupy less preferred types of habitat, and one may guess that in these the death rate from starvation, predation or other factors, will be higher than normal. Except in so far as surplus individuals may be able to come in and replace any successful territory-holders who have died, they are doomed to a sterile, and probably short, life.

It has been for some years unfashionable to think that the numbers of birds are limited in this way, but there is strong general evidence that it is the most important control on many species of land-birds.[45, 253] Counts over 8 to 20 years in five American woods of different types showed that populations were least variable in habitats which were optimal for the species. In the Netherlands also, the great tit occupies coniferous woodland only after the available deciduous woods are full, so that while the numbers in the former fluctuate from year to year, those in the latter are nearly constant.[190] Whenever a new area of a given type of habitat is made by man, it is rapidly occupied by appropriate birds, and no one has ever shown, or even suggested, that there is an accompanying fall in population in the previously-existing habitats. There must, therefore, have been a surplus population which would have died without offspring but for man's action in providing it with somewhere to live.

Such an increase in numbers is shown most spectacularly by the Forestry Commission's plantations. These are mostly of alien conifers, and are not likely to be the preferred habitat for many British birds. Nevertheless they will serve, although it has been shown, both in England and in the Netherlands, that the great tit does not breed so successfully in them as in deciduous woods. The increase in numbers of such birds as tits, blackbirds, robins, hedge-sparrows (*Prunella vulgaris*), willow-warblers, chaffinches and tree-pipits must have been immense. At the same time, since most of the plantations are made on open ground, there has been a concomitant fall in the numbers of meadow-pipits and skylarks, and probably of cuckoos, curlews and a few other species. There are further changes as the trees grow and are thinned; species such as hedge-sparrow and whitethroat, which need thick cover, disappear, and others, such as coal tit and chaffinch, become more important.[406]

Observations of the bearded tit (*Panurus biarmicus*) show how this new occupancy may arise. The bird nests, and lives throughout the year, exclusively in reed beds and similar vegetation. It was formerly found in such places over most of the south of England, but declined greatly in the first part of this century, and was further reduced by the hard winter of 1947. It was then almost extinct, and only two breeding pairs were known in Suffolk and one solitary male in Norfolk.

There was a gradual increase in the population, and in 1959, after a very successful breeding season, bearded tits were seen in various parts of England for the first time for some years, and in this and the following years many counties produced their first records.[20] There were reports, for example, from Warwickshire in 1959, 1964 and 1966, Staffordshire in 1963, 1965 and 1966, and Worcestershire in 1965 and 1966. All these were in autumn or winter, and breeding was not known to occur. In 1967, however, a single male was present at what might well be a suitable breeding place in Worcestershire throughout the year.

That these winter birds were moving about is shown by the fact that two birds ringed in Dorset in October were caught a few weeks later 60 miles away in Somerset. In 1965 some of these vagrants came from the continent, where also there has been an increase in numbers.

The birds that cannot find suitable territory in their old home are likely to be the young, and there is evidence that in the American redstart (*Setophaga ruticilla*) this is so.[110] Fewer first-year males breed in deciduous forests, which are the preferred habitat, and more in suboptimal habitats.

It has now been shown fairly convincingly that the population of red grouse in any year is determined by the number of cock birds holding territories on suitable moorland in the previous autumn.[383] Other factors, including predation, shooting and bad weather in the intervening winter, remove only the surplus birds that would not breed anyway, or, if they did, would be unsuccessful because they held territories on unsuitable land. When birds were removed from a territory their places were quickly taken by other birds, so that the total breeding population remained constant.[385] The ptarmigan appears to be controlled in a similar way.[384]

In captivity, chicks from eggs collected from good moors, that is those on which breeding is known to be highly successful, survive better than those from poor moors.[173] This must mean either that the difference is hereditary, or that it comes in through the conditions under which the parents live; it certainly cannot be due to the better food supply for the chicks on the good moors, since in captivity both groups are treated alike. Good moors are, in part at least, those on base-rich soils.[261] A similar type of territorial control seems to apply to the dunlin (*Calidris alpina*) in Finland.[331]

The provision of nest-boxes shows even more clearly how a small alteration in the habitat may allow more birds to survive. There is no likelihood that the large new population of the pied flycatcher in the Forest of Dean[56] (Table 8.2), which has now been paralleled on a smaller scale in other woods in the west of England, has led to any reduction in the numbers of the bird in its former rather restricted haunts. The same applies, less spectacularly, to the increase in numbers of great and blue tit that must have taken place with the erection of nest-boxes in many woods and gardens, and to the appearance of colonies of sand-martins in the first year in which gravel digging has exposed a new sandy cliff. Most of the observations on populations of tits, which have been claimed to show the overriding importance of food, are uncontrolled experiments in that they have been made in heavily boxed areas with a population density almost certainly higher than normal. Even under similar circumstances to these, crowding rather than food may be the limiting factor; a population of wood ducks (*Aix sponsa*) induced by the provision of nest-boxes became self-limiting, the production of young per pair being inversely proportional to density apparently because of interference at the nest-site.[176]

9.5 HABITAT SELECTION[90, 161]

Just as man, in buying a house, takes many things into account but in the end chooses (in large part unconsciously) one that in total comes nearest to some sort of ideal image of the perfect place in which to live, so it seems birds choose (by instinct, some would say) the nearest they can come to an ideal. This would be the place where they can nest without disturbance, where there is plenty of food, where the climate is never harsh, and where there are no predators. Individuals that had the hereditary endowment to cause them to make such a choice would leave the most offspring, and so would be selected, and their genes would spread through the population.

In practice, birds, like man, must often settle for less than the ideal. If most of the possible conditions are good enough, selection may go by the presence or absence of one condition only. The tree-pipits that invade young coniferous plantations or regenerating woodland (Chapter 7) and sing from trees only a few feet high, are not getting the best situation, but making do with something better than nothing. They are likely, in accordance with what has been said above, to be young birds.

Birds, then, probably choose what may be called 'total habitat', but may be influenced by one factor more than another at different times. Some of the important factors are known, such as nest-sites and song-posts; food must be adequate; climate must be suitable. It is possible to show that the limits of distribution of some species follow particular climatic lines; that of the nightingale in England for example, is close to the area enclosed within the 66°F (18·9°C) isotherm[275] (Fig. 9.5). The northern limit of wintering of many east European birds is the January isotherm of −6°C.[341] Some introduced birds have been shown to be successful only where the climographs of their new country in the breeding season fall within those of their home.

Sometimes it is not easy to see the common features of the various places where a bird lives. The fieldfare (*Turdus pilaris*) is characteristically a bird of the taiga, breeding in coniferous and birch woods in Siberia and Scandinavia to about 71°N, or to the July isotherm of 50°F (10°C). In this century it has invaded central Europe, and breeds for example in Switzerland in a marshy valley with clumps of deciduous trees. It is here found alongside the marsh-warbler, a southern European bird limited to the north by the July isotherm of 62°F (18·9°C), which has been extending its range northwards at the same time as the fieldfare has been moving south.

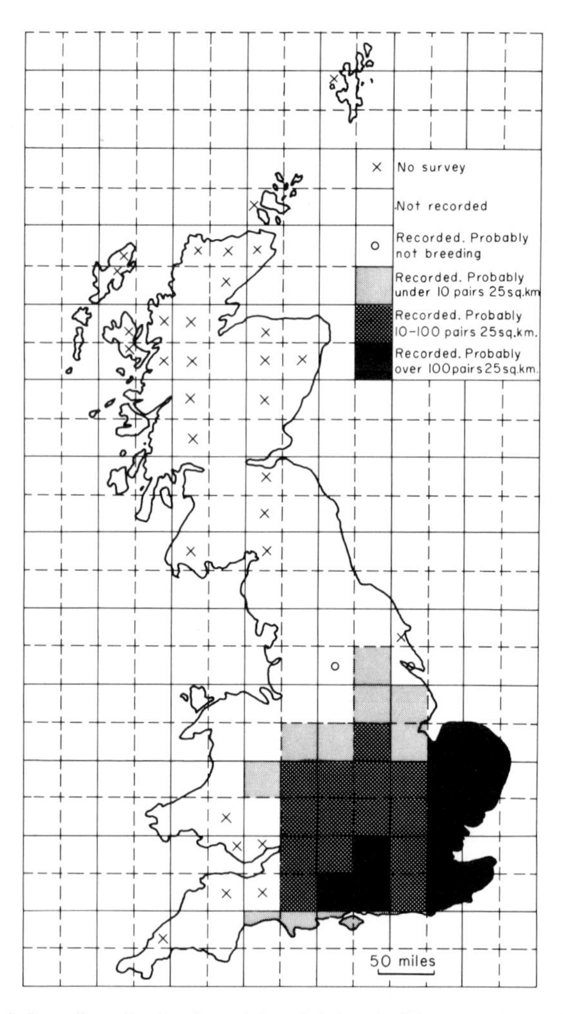

Fig. 9.5 (**a**) Breeding distribution of the nightingale. The map shows the highest density estimated by experienced observers in any one 5×5 km square within each 50×50 km square of the national grid. The west Gloucestershire and Somerset squares would probably have scored under 10 or between 10 and 100 pairs had there been any observers. (From C. A. Norris, 1960, Fig. 3, *Bird Study* **7**, 134.)

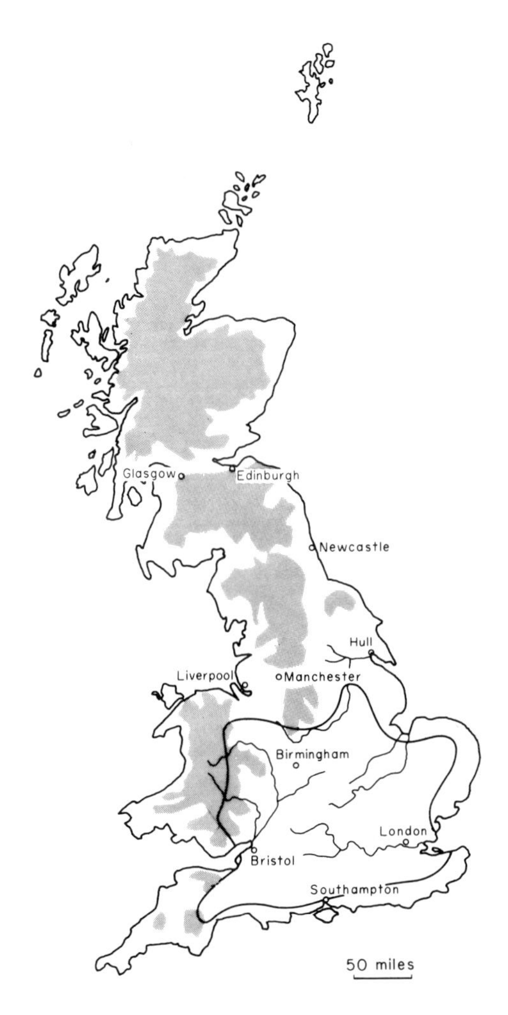

(**b**) Mean daily maximum temperature, June; 66°F isotherm. The cooler east and southeast coasts, the only parts outside the isotherm that have high densities of nightingales, have high minimum temperatures in June. (From W. B. Yapp, 1962, *Birds and Woods*, Figures 15 and 16, Oxford University Press, London.)

9.51 The influence of other species

Amongst the features of the habitat that may influence a species in its choice is the presence or absence of other birds. In this connection much play has been made in recent English writing on ornithology with Gause's hypothesis,[121, 122] of which it need only be said that something very like it was enunciated 30 years before Gause by Grinnell,[140] that Gause makes it clear (as his followers do not) that he was concerned with organisms living in a microcosm, not in their natural habitats, and that many of the views attributed to Gause cannot be found in his writings. When he is correctly quoted as saying 'as a result of competition, two similar species scarcely ever occupy similar niches' he is saying something that is not true, unless 'niche' is given the tautologous meaning (as by implication it sometimes is) 'the habitat exclusively occupied by one species'. There are plenty of examples of closely-related species living in similar environments; blue tit and great tit, song-thrush and blackbird, willow-warbler and wood-warbler, pied and spotted flycatcher are pairs that can all be found together in the same wood. One of the most striking European examples is given by the congeneric blackcap and garden-warbler; the females are almost indistinguishable, and the songs of the males sometimes so. The ranges of the two are almost the same, the garden-warbler extending a little further to the north and the blackcap a little further to the south. In England they may be found in the same woods, and breeding within a few yards of each other in what is, so far as man can see, a continuous type of vegetation. Closely-related birds are not merely in similar habitats, they are (except in the sense that no two individuals can live in the same place) in the same one. Another good example is given by the herring-gull and the lesser black-backed gull.[50] They are merely the extremes of a graded series of colour-forms found all round the northern hemisphere, in which each step justifies only subspecific separation, but in which the extremes are regarded as good species. On Walney Island (Lancashire) they nest together at high density and do not normally interbreed. Occasional crosses are fertile.

The members of such species-pairs may show differences in the way in which they use the habitat that they share; they may have different preferences in nest-sites, or take different proportions of the available foods. Since they sometimes also make their nests in exactly similar places, or feed on the same species of insect or seed, they nevertheless compete with each other at times, and should be contrasted, rather than compared, with the micro-organisms living in test-tubes with which Gause experimented.

If closely-related species did not live in similar habitats it is difficult to see how speciation could have taken place, for when one species has split into two there must have been gradual divergence of structure. Grinnell, in discussing the probable origin of the chestnut-backed chickadee (*Parus rufescens*) of the Pacific Coast of North America, from the boreal or brown-capped chickadee (*P. hudsonicus*), which ranges across the continent, together with their present distributions, came to the conclusion that one species could split into two only if the population became divided into two parts, and that two similar species, of recent common ancestry, would only be found together if one of them had invaded the area of the other after the two had become different enough not to interbreed.[140] It should be noted that the necessity of isolation for speciation was thus put forward on ecological grounds long before the geneticists postulated the same thing by mathematics. Isolation, which for most ornithologists means geographical isolation, is now accepted as an essential prerequisite for the splitting of a species. The arguments for it are strong, but it is difficult to see where the isolation that has produced the rich avifauna of the tropics has taken place. Other types of isolation, based mainly on habits or sexual selection, need investigation. There is some statistical evidence that in a given type of habitat there are more, not fewer, congeneric species than would be expected by chance, which suggests speciation *in situ*.[396]

Although closely-related species often live together there are many recorded instances where the presence of one species is said to inhibit that of a relative. It is said for example, that if the house wren (*Troglodytes aedon*) invades an area, Bewick's wren (*T. bewickii*) leaves. It may, however, come back.[48]

In the Cape Verde Isles the barn owl is the only owl, and has a wider range of habitats than usual; it searches for food by quartering the open country, like the short-eared owl elsewhere.[40] This habit, however, is sometimes, though not regularly, seen in England, where the range overlaps that of the short-eared owl.

The chaffinches on the Canary Islands have been referred to in Chapter 3. Some other examples also must be taken with caution. On islands in the Baltic the coal tit is said to breed in deciduous woods as well as coniferous ones because marsh and willow tits are absent.[352] But it breeds in some deciduous woods where one or other of these species is present in England and Wales, presumably because they provide suitable nesting-sites—cracks near the ground, rather than the straightforward holes used by most tits.

There are many recorded examples of one species driving another from its territory, or more often from the vicinity of the nest, and such attacks seem to be just as frequent, or more so, between species that are not closely related as between those that are.[406] They probably have some effect in spacing the birds through the habitat but little on excluding a species. In contrast, many species are very tolerant of others, and feed and sing in the same tree at the same time or in quick succession. Where the ranges of the chaffinch and brambling overlap, the birds do not interfere with each other, and their territories overlap.[377] The habitats of meadow-pipit (*Anthus pratensis*) and tree-pipit are usually distinct, but in young plantations, and also near the tree limit, both altitudinal and latitudinal, the tree-pipit may use very small trees and the meadow-pipit may sing from trees or bushes and the two do not interfere with each other.

9.52 Changing habitat

Although one must in general assume that every species has a characteristic or preferred habitat, there are many examples, some of which have been mentioned already, of birds living in uncharacteristic places or even apparently, in human terms, changing their habitat. The whinchat (*Saxicola rubetra*) may sing from oaks 30 or 40 feet high, although bushes of three or four feet, which it normally prefers, are nearby. Magpie (*Pica pica*), wren and blackbird, normally woodland birds, can all be found where there are no trees, especially on hills and in the north. Trees as such are apparently unnecessary, but usually their surroundings provide the things that are. Beyond the tree limit in Finmark, where there are no twigs, the magpie makes nests of wire.[86] The reed bunting (*Emberiza schoeniclus*), whose usual habitat is scrubby streamsides, has in recent years been seen more and more in drier bushy regions, more typical of its relative the yellowhammer.[183]

The most striking changes are those to man-made habitats. Few people have ever seen nests of the house-martin except under the eaves of houses, but there are still some under overhanging cliffs in Yorkshire and elsewhere. The swallow hardly ever builds its nest now except inside barns and such places. The sand-martin still usually makes its own tunnels in sandy cliffs, but will use drain holes in a vertical wall. The common nighthawk breeds naturally on gravel banks and such places, but often uses the flat roof of modern buildings.

No species of bird has become completely dependent on man in the sense of living wholly in an entirely man-made environment, but many individuals of several species live in suburban or even urban surroundings. One may guess that the change in habit began with birds of the year being unable to find territory in their normal habitat, and by the return of their young to the place where they were bred, or to similar surroundings.

House-sparrows have long been known as commensals with man, and many species, such as blackbirds, have long been common in gardens, which are a sort of modified woodland. More recently other species have invaded gardens and town parks. Green woodpecker (*Picus viridis*), great spotted woodpecker, jay and wood-pigeon have come into parks in London and elsewhere in this century.

In this sort of adaptation there is much geographical variation. Pied flycatchers nest in towns (provided there are nest-boxes) in Sweden and Finland, but not in England. The tree-sparrow (*Passer montanus*) is a town bird in parts of the continent, taking the place of the house-sparrow elsewhere. Robins are found only in secluded woodlands over most of Europe. There is some evidence that garden birds are more successful than forest birds both in temperate and in tropical regions, perhaps because of the absence of predators.[323] Whatever the reason there can be no doubt that birds are a successful and active group, able to learn to take advantage of changing conditions and new opportunities.

9.6 POPULATION CONTROL: SUMMARY[61, 62]

The most plausible hypothesis of population control is that it depends on many factors, but that at any one time and under any one set of circumstances, one of them is limiting, in the sense that a change in it has an immediate effect on the population while the others can alter to some extent without affecting population at all. If one of these other factors changes enough in the right direction it will in turn become limiting. Nesting-sites have clearly in the past limited the population of pied flycatchers in the Forest of Dean, but since 1951 the fluctuations in its numbers must have been due to something else, since not all the boxes have been occupied. Something other than nesting-sites must have limited the numbers of the fulmar in the past, since the cliffs on which they now nest have always been there and have not altered.

The way in which factors interact may be illustrated by some observations made in woodlands in Czechoslovakia.[374] A heavy outbreak of infestation by the gypsy-moth (*Liparis dispar*) doubles the bird population, but the defoliation caused by the caterpillars exposes chicks in the nest both to bad weather and to greater predation than usual, so that the proportion of nestlings that fly is lowered. Hence there is only a temporary fluctuation in the population. Another type of interaction may occur as a wood matures and the canopy increases. The ground vegetation decreases and hence the humus does too, and also the invertebrates living in the soil. There is thus less food for ovenbirds (*Seiurus aurocapillus*) and the size of their territories increases as their population falls. Are their numbers regulated by territory, by food, or by the vegetation?[343]

Birds near the edge of their range, and probably all migratory birds, are subject to occasional heavy mortality by extreme weather conditions. In exceptionally cold weather, such as the winters of 1916–17, 1928–29, 1939–40, 1946–47, 1961–62 and 1962–63, the numbers of nearly all British birds are reduced, and some, such as the wren and kingfisher, practically disappear. Most recover within one season, but those that have been most severely reduced take longer. After the unusual event of two cold winters running, in 1961–63, the wren did not recover until 1966 or 1967, while the kingfisher is still, in 1969, below normal. This effect was very well shown by the counts of the heron, whose fluctuations have exactly followed the hard winters.[339] For such birds weather is the limiting factor; if the climate permanently improved, something else would take its place, and presumably at present does so in less cold parts of Europe. The general importance of the winter in controlling population is shown by the fact that the numbers of residents and summer visitors in the Dean nest-boxes fluctuate independently.

Appendix 1

A classification of birds and list of species mentioned in the text

This classification is, except in one respect, that of Wetmore,[394] which is a development of that of Gadow. For reasons given in Chapter 3 I have associated the main groups of flightless birds as the superorder Ratitae, so agreeing with Gadow but not Wetmore. Suborders are given only where an order is divided into two or more easily-recognized groups, but where given are complete. Families (included only for the Oscines), genera and species are only those mentioned in the text, except that where the sense of the text requires a reference only to a family or a genus, a characteristic species is generally given in this classification; they are arranged for convenience in alphabetical order.

The classification in *A New Dictionary of Birds*[10] differs chiefly in its treatment of the order Passeriformes. The buntings and some others are split off from the Fringillidae as a family of their own, the Emberizidae; the family Ploceidae is divided into Ploceidae and Estrildidae; and the Regulidae, Sylviidae, and Turdidae are merged in the Muscicapidae.

CLASS AVES: Birds

Subclass Archaeornithes (= Saurornithes)
Archaeopteryx lithographica only
Subclass Neornithes
Superorder Odontognathae
Order Hesperornithiformes
Hesperornis

Order Ichthyornithiformes

Ichthyornis

Superorder Impennes, penguins

Order Sphenisciformes

Aptenodytes forsteri, emperor penguin

A. patagonica, king penguin

Pygoscelis adeliae, Adélie penguin

Superorder Ratitae

Order Struthioniformes

Struthio camelus, ostrich

Order Rheiformes

Pterocnemia pennata, Darwin's rhea

Rhea americana, common rhea

Order Casuariiformes

Casuarius casuarius, Australian cassowary

Order Aepyornithiformes

Aepyornis, elephant-bird

Order Dinornithiformes

Dinornis, moa

Order Apterygiformes

Dromaius novaehollandiae, emu

Apteryx australis, brown kiwi

Order Tinamiformes, tinamous

Crypturellus soui, pileated tinamou

Superorder Carinatae

Order Gaviiformes, loons or divers

Order Podicipediformes, grebes

Centropelma micropterum, flightless grebe

Podiceps cristatus, great crested grebe

Order Procellariiformes

Diomedea epomophora, Royal albatross

D. immutabilis, Laysan albatross

Fulmar glacialis, fulmar

Puffinus puffinus, Manx shearwater

P. tenuirostris, short-tailed shearwater

Order Pelecaniformes

Fregata, frigate birds

Nannopterum harrisi, flightless cormorant

Pelecanus occidentalis, brown pelican

Phaeton aethereus, red-billed tropic bird

P. lepturus, yellow-billed tropic bird

Phalacrocorax carbo, common cormorant

Sula bassana, gannet

Order Ciconiiformes
Ardea cinerea, common or grey heron
Ciconia ciconia, white stork
Order Anseriformes
Suborder Anhimae, screamers
Chauna torquata, crested screamer
Suborder Anseres, ducks, geese, and swans
Aix sponsa, wood duck
Anas platyrhynchos, wild duck or mallard
Tachyeres, steamer ducks
Tadorna tadorna, shelduck
Order Falconiformes
Suborder Cathartae, American vultures
Cathartes aura, turkey vulture
Gymnogyps californianus, Californian condor
Vultur gryphus, Andean condor
Suborder Falcones, hawks and old-world vultures
Accipiter nisus, sparrow-hawk
Buteo buteo, common buzzard
Aquila chrysaetos, golden eagle
Falco aesalon, merlin
F. peregrinus, peregrine falcon
F. subbuteo, hobby
F. tinnunculus, kestrel
Pandion haliaetus, osprey
Rostrhamus sociabilis, Everglade kite
Order Galliformes
Centrocercus urophasianus, sage grouse
Coturnix coturnix, quail
Lagopus lagopus, grouse
L. mutus, ptarmigan
Lyrurus tetrix, blackcock and greyhen
Meleagris gallopavo, turkey
Pavo cristatus, peafowl
Perdix perdix, partridge
Phasianus colchicus, pheasant
Syrmaticus reevesii, Reeves' pheasant
Opisthocomus hoazin, hoatzin
Order Gruiformes
Family Gruidae, cranes
Grus americana, whooping crane
Family Otididae, bustards

Family Rallidae, rails
Atlantisia rogersi, flightless rail
Crex crex, corncrake
Fulica atra, coot
Gallinula chloropus, moorhen
Rallus madagascarensis, Madagascar rail
Order Charadriiformes
Suborder Charadrii, waders
Calidris alpina, dunlin
Capella gallinago, common snipe
Charadrius hiaticula, ringed plover
Haematopus ostralagus, oyster-catcher
Numenius arquata, curlew
Phalaropus fulicarius, grey (= red) phalarope
Philomachus pugnax, ruff and reeve
Pluvialis dominica, American golden plover
Recurvirostra avosetta, avocet
Scolopax rusticola, woodcock
Vanellus vanellus, lapwing
Suborder Lari, gulls
Larus argentatus, herring-gull
L. fuscus, lesser black-backed gull
L. hyperboreus, glaucous gull
L. marinus, great black-backed gull
L. occidentalis, western gull
L. ridibundus, black-headed gull
Rissa tridactyla, kittiwake
Sterna fuscata, sooty tern
Suborder Alcae, auks
Alca torda, razorbill
Fratercula arctica, puffin
Pinguinus impennis, great auk
Uria alge, guillemot
Order Columbiformes
Columba oenas, stock-dove
C. palumbus, wood-pigeon
Ectopistes migratorius, passenger pigeon
Raphus cucullatus, dodo
Streptopelia risoria, ringed turtle dove
Syrrhaptes paradoxus, Pallas's sand-grouse
Zenaidura macroura, mourning dove
Order Psittaciformes
Melopsittacus undulatus, budgerigar

Psittacus erithacus, grey parrot
Strigops habroptilus, kakapo
Order Cuculiformes
Cuculus canorus, cuckoo
Order Strigiformes
Asio flammeus, short-eared owl
A. otus, long-eared owl
Bubo virginianus, great horned owl
Strix aluco, tawny owl
Tyto alba, barn owl
Order Caprimulgiformes, nightjars
Chordeiles minor, common nighthawk
Phaelaenoptilus nuttali, poorwill
Steatornis capensis, oilbird
Order Apodiformes
Suborder Apodi
Apus apus, common swift
A. melba, Alpine swift
Collocalia vestita, white-nest swiftlet
Suborder Trochili
Archilochus alexandri, black-chinned humming-bird
A. colubris, ruby-throated humming-bird
Calypte costae, Costa's humming-bird
Eugenes fulgens, Rivoli's humming-bird
Lampornis clemencii, blue-throated humming-bird
Mellisuga helenae, bee humming-bird
Order Coliiformes
Colius colius, white-backed mousebird
Order Trogoniformes, trogons
Order Coraciiformes
Alcedo atthis, common kingfisher
Buceros bicornis, great hornbill
Merops bullocki, red-throated bee-eater
Order Piciformes
Dendrocopos major, great spotted woodpecker
D. minor, lesser spotted woodpecker
Indicator indicator, black-throated honey-guide
Jynx torquilla, wryneck
Picoides tridactylus, three-toed woodpecker
Picus viridis, green woodpecker
Order Passeriformes
Suborder Eurylaimi, broadbills
Suborder Tyranni

Chiroxiphia pareola, manakin
Empidonax traillii, alder flycatcher
Suborder Menurae, lyrebirds
Suborder Oscines (= Passeres)
Family Alaudidae
Alauda arvensis, skylark
Family Certhiidae
Certhia familiaris, tree-creeper
Family Cinclidae
Cinclus cinclus, dipper
Family Corvidae
Corvus brachyrhynchos, (American) common crow
C. corax, raven
C. corone corone, carrion crow
C. corone cornix, hooded crow
C. frugilegus, rook
C. monedula, jackdaw
Garrulus glandarius, jay
Nucifraga caryocatactes, nutcracker
Perisoreus canadensis, Canada jay
Pica pica, magpie
Family Fringillidae
Cactospiza (*Camarhynchus*) *pallida*, Galapagos woodpecker finch
Carduelis carduelis, goldfinch
C. flammea, redpoll
Chloris chloris, greenfinch
Coccothraustes coccothraustes, hawfinch
Emberiza calandra, corn bunting
E. citrinella, yellowhammer
E. hortulana, ortolan
E. schoeniclus, reed bunting
Fringilla coelebs, chaffinch
F. (*Montifringilla*) *montifringilla*, brambling
F. teydea, blue chaffinch
Junco hyemalis, slate-colored junco
Loxia curvirostra, crossbill
Melospiza lincolnii, Lincoln's sparrow
M. melodia, song-sparrow
Passerculus sandvichensis, savannah sparrow
Passerella iliaca, fox sparrow
Pooecetes gramineus, vesper sparrow
Pyrrhula pyrrhula, bullfinch

Pyrrhuloxia (Richmondena) cardinalis, cardinal
Serinus canaria, canary
Spizella arborea, (American) tree-sparrow
Zonotrichia albicollis, white-throated sparrow
Z. leucophrys, white-crowned sparrow

Family Hirundinidae
Delichon urbica, house-martin
Hirundo rustica, (barn) swallow
Petrochelidon pyrrhonota, cliff swallow
Riparia riparia, sand-martin

Family Icteridae
Agelaius phoeniceus, red-winged blackbird
A. tricolor, tricolored redwing
Dolichonyx orizivorus, bobolink
Molothrus ater, brown-headed cowbird
M. badius, bay-winged cowbird
Scaphidura oryzivora, giant cowbird

Family Laniidae
Lanius collurio, red-backed shrike

Family Mimidae
Mimus polyglottus, mocking bird
Toxostoma rufum, brown thrasher

Family Motacillidae
Anthus pratensis, meadow-pipit
A. trivialis, tree-pipit
Motacilla aguimp, African pied wagtail
M. flava, yellow wagtail

Family Muscicapidae
Muscicapa (Ficedula) albicollis, white-collared flycatcher
M. (F.) hypoleuca, pied flycatcher
M. striata, spotted flycatcher

Family Nectariniidae
Cinnyris coccinigaster, splendid sunbird

Family Paridae
Aegithalos caudatus, long-tailed tit
Parus ater, coal tit
P. atricapillus, black-capped chickadee
P. caeruleus, blue tit
P. carolinensis, Carolina chickadee
P. cinctus, Siberian tit (grey-headed chickadee)
P. cristatus, crested tit
P. hudsonicus, brown-capped chickadee
P. inornatus, plain tit

P. major, great tit
P. montanus, willow tit
P. palustris, marsh tit
P. rufescens, chestnut-backed chickadee
Psaltriperus minimus, common bush-tit
Family Paradoxornithidae
Panurus biarmicus, bearded tit
Family Parulidae (= Compsothlypidae)
Dendroica kirtlandii, Kirtland's warbler
Seiurus aurocapillus, ovenbird
Setophaga ruticilla, American redstart
Family Ploceidae
Estrilda troglodytes, black-rumped waxbill
Hypochera chalybeate, Senegal combassou
Lagonostricta senegala, Senegal fire-finch
Passer domesticus, house-sparrow (English Sparrow)
P. montanus, (European) tree-sparrow
Quelea quelea, red-billed dioch
Steganura (*Vidua*) *paradisaea*, paradise whydah
Taeniopygia castanotis, zebra finch
Family Prunellidae
Prunella vulgaris, hedge-sparrow
Family Ptilonorhynchidae
Ptilonorhynchus violaceus, satin bower-bird
Family Regulidae
Regulus regulus, goldcrest
Family Sittidae
Sitta europaea, nuthatch
S. pusilla, brown-headed nuthatch
Family Sturnidae
Sturnus vulgaris, starling
Family Sylviidae
Acrocephalus palustris, marsh-warbler
A. schoenobanus, sedge-warbler
A. scirpaceus, reed-warbler
Locustella naevia, grasshopper warbler
Phylloscopus collybita, chiffchaff
P. sibilatrix, wood-warbler
P. trochilus, willow-warbler
Sylvia atricapilla, blackcap
S. borin, garden-warbler
S. communis, whitethroat
S. undata, Dartford warbler

Family Troglodytidae
Campylorhynchus brunneicapillus, cactus wren
Troglodytes aedon, house wren
T. bewickii, Bewick's wren
T. troglodytes, wren
Family Turdidae
Cyanosylvia suecica, bluethroat
Erithacus rubecula, (European) robin
Luscinia megarhynchos, nightingale
Oenanthe oenanthe, wheatear
Phoenicurus phoenicurus, redstart
Saxicola rubetra, whinchat
S. torquata, stonechat
Turdus merula, (European) blackbird
T. musicus, (European) redwing
T. philomelos, song-thrush
T. pilaris, fieldfare
T. viscivorus, mistle-thrush

Appendix 2

Glossary

Active transport any process by which a substance in solution is conveyed through a membrane more rapidly than it would be by simple diffusion, difference of pressure or other simple physical process.

Amphicoelous adjective applied to a vertebra whose centrum is concave both fore and aft.

Amphiplatyan adjective applied to a vertebra whose centrum is flat both fore and aft.

Autolysis the digestion or solution of any once living material by enzymes contained within itself.

Convergence the evolution of two groups of organisms, or of the corresponding parts of the body in such groups, so that they come to resemble one another.

Cretaceous one of the geological periods, see Table A1.

Cryptic adjective applied (1) to an animal that resembles its background, or a particular feature in the background, so that it is difficult to see, (2) to the coloration that produces such resemblance.

Dimorphism the existence of two different forms of the same species: especially *Sexual dimorphism*, where male and female differ in characters other than those directly connected with sex.

Diopter a measure of the extent to which the lens of the eye can accommodate so that a clear image of objects close to the eye is formed on the retina. The accommodation in diopters is the reciprocal of the focal length in metres of an accessory lens (as in spectacles) that would produce the same effect as does the change in shape of the lens in the living eye.

Distal further from the centre of the body, or from the point of origin of a limb or other projecting structure.

Endemic occurring naturally only in the place stated; (the word has a different meaning in pathology).

Eocene one of the geological periods; see Table A1.

Frugivorous feeding on fruit.

Gular pertaining to the throat.

Hamuli small hooks on the anterior barbules of a feather.

Hertz (Hz) A frequency of one per second.

Heterocoelous adjective applied to a vertebra whose centrum has surfaces that are concave in one direction and convex in another at right angles to the first; saddle-shaped.

Homogametic possessing two similar sex-chromosomes, one derived from each parent; homozygous for sex.

Homologous adjective applied to two or more structures that are considered to be derived by the evolution of a single structure that was present in a common ancestor.

Homozygous possessing two hereditary factors or genes for a given character, one derived from each parent.

Hyperdactyly The presence of more than five digits in a pentadactyl limb.

Hyperglycaemia the state of having more than the usual amount of sugar in the blood.

Hyperphalangy the presence of an abnormally large number of phalanges (joints) in the digits of a pentadactyl limb.

Jurassic one of the geological periods; see Table A1.

Kinetism the condition where the skull is so constructed that the facial portion, or part of it, can be moved forward and upward in relation to the rest. It is found in different forms in some fishes and reptiles (especially snakes), and in ratites, parrots and some other birds.

Mitochondria small structures in the cell with folded membranous walls, the site especially of enzymes concerned with oxidation.

Monophyletic implies that common ancestors for all the members of a taxon can be postulated such that they would be included in the same taxon, or in another taxon of the same rank with no intermediates that would be put in another taxon of that rank; in the following diagrams orders P and R are monophyletic.

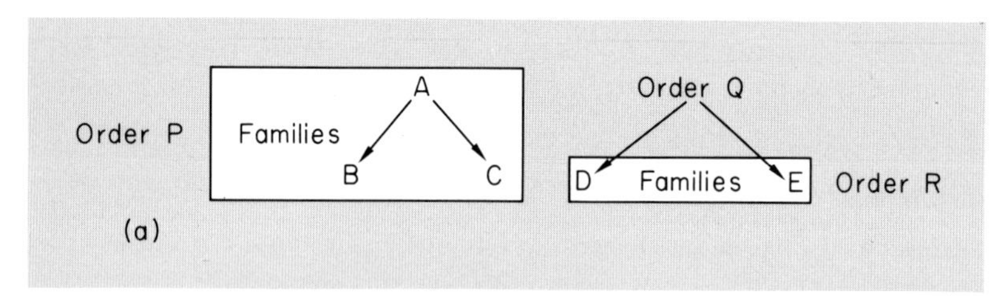

Polyphyletic (and *Diphyletic*) are the opposite of this, and imply that the nearest common ancestors lie so far back that they would be placed in a different taxon of the same rank, with other taxa of this rank intervening; diphyletic is used where there are two taxa in parallel in the intervening stage, polyphyletic where there are more than two, thus

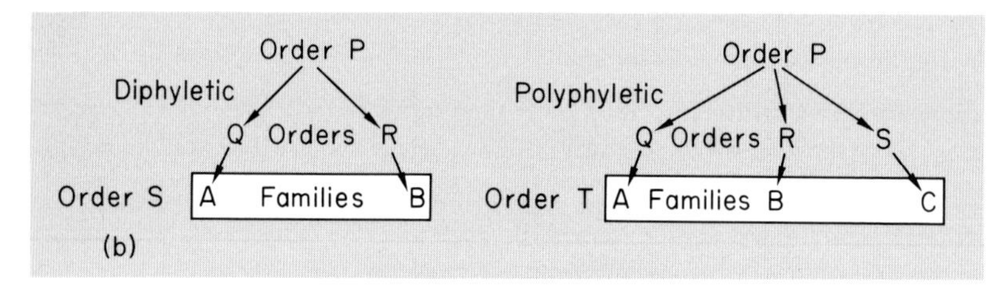

Where such a descent is well-established, the final taxa (S in the diphyletic diagram, T in the polyphyletic) should be broken up, although their old names often remain in general use.
(The words diphyletic and polyphyletic are often incorrectly used to imply merely that a group has two or more well-marked divisions, as in order P above.)

Neopallium a part of the forebrain, slightly developed in reptiles, but making most of the cerebral hemispheres in mammals.

Ornithine cycle a series of chemical processes in which ammonia, derived from proteins, is converted into urea. The aminoacid ornithine enters into an early stage of the reaction and is regenerated at the end, hence the name cycle.

Pentadactyl literally, having five fingers, but applied not only to the hypothetical limb of the first terrestrial vertebrates, which probably had five fingers (or toes), but to any limb derived from this irrespective of the number of digits.

Permian one of the geological periods; see Table A1.

Pliocene One of the geological periods; see Table A1.

Pollex the first digit of the forelimb or arm; in man, the thumb.

Polyandry the simultaneous association of one female with two or more male breeding partners.

Polygyny the simultaneous association of one male with two or more female breeding partners; usually called polygamy, but this term strictly includes polyandry as well.

Polyphyletic see monophyletic.

Proximal nearer to the centre of the body, or to the point of origin of a limb or other projecting structure.

Pterylosis the pattern of arrangement of feathers in tracts or pterylae.

Remiges (sing. *Remex*) the primary and secondary feathers of the wing.

Taiga the region of natural coniferous forest of the northern hemisphere.

Taxon any classificatory group, from a subspecies to a kingdom; plural, taxa.

Taxonomy the theory, study or arrangement of taxa; the science of classifying.

Triassic one of the geological periods; see Table A1.

Table A1. Geological periods and the evolution of birds and mammals. The horizontal divisions show the duration of each period, and the figures show the time in millions of years that has elapsed since the beginning of each.

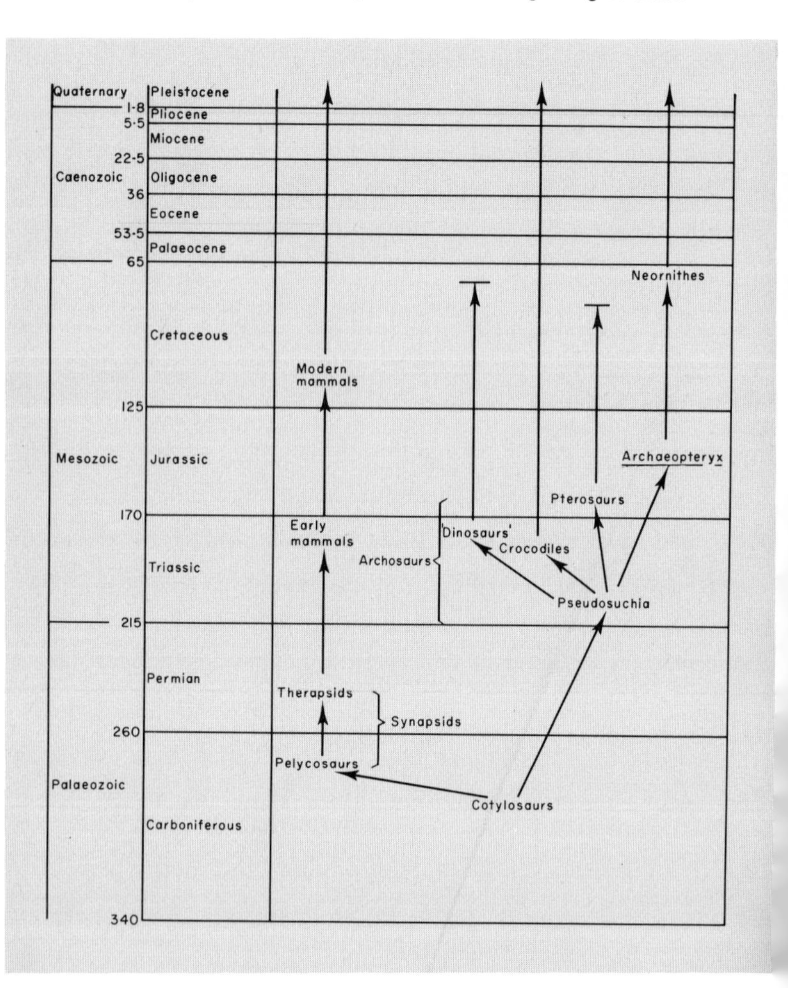

References

GENERAL

1. BEDDARD, F. E. (1898). *The Structure and Classification of Birds.* Longmans, Green, London.
2. EVANS, H. A. (1899). Birds. In *Cambridge Natural History*, ed. HARMER, S. F. and SHIPLEY, A. E., Macmillan, London.
3. GRASSÉ, P. P., ed. (1950). *Traité de Zoologie, 15, Oiseaux.* Masson, Paris.
4. MARSHALL, A. J., ed. (1960–61). *Biology and Comparative physiology of Birds.* Academic Press, New York and London.
5. NEWTON, A. (1896). *A Dictionary of Birds.* A. and C. Black, London.
6. POUGH, R. H. (1949), *Audubon Land Bird Guide.* Doubleday, Garden City, New York.
7. POUGH, R. H. (1951). *Audubon Water Bird Guide.* Doubleday, Garden City, New York.
8. POUGH, R. H. (1957). *Audubon Western Bird Guide.* Doubleday, Garden City, New York.
9. STURKIE, P. D. (1965). *Avian Physiology*, 2nd edn. Baillière, Tindall and Cassell, London.
10. THOMSON, A. L. ed. (1964). *A New Dictionary of Birds.* Nelson, London.
11. WITHERBY, H. F., ed. (1941). *The Handbook of British Birds.* Witherby, London.
12. WORDEN, A. N. (1958). *Functional Anatomy of Birds.* Cage birds. Iliffe, London.
13. YAPP, W. B. (1965). *Vertebrates, their Structure and Life.* Oxford University Press, New York.
14. YARRELL, W. (1871–85). *A history of British birds*, 4th edn. Van Voorst, London.

SPECIAL REFERENCES

15. AKESTER, A. R. (1960). The comparative anatomy of the respiratory pathways in the domestic fowl (*Gallus domesticus*), pigeon (*Columba livia*) and domestic duck (*Anas platyrhynca*). *J. Anat.* **94**, 487–505.

16. ANDREW, R. J. (1956). Territorial behaviour of the yellowhammer *Emberiza citrinella* and corn bunting *E. calandra*. *Ibis* **98**, 502–5.
16a. ANON. (1968). *NERC A. Rep. 1967–68* 66.
17. ARMSTRONG, E. A. (1956). Territory in the wren *Troglodytes troglodytes*. *Ibis* **98**, 430–7.
18. ARMSTRONG, E. A. (1963). *A Study of Bird Song*. Oxford University Press, London.
19. AUBER, L. (1957). The distribution of structural colours and unusual pigments in the class Aves. *Ibis* **99**, 463–76.
20. AXELL, H. E. (1966). Eruptions of bearded tits during 1959–65. *Br. Birds* **59**, 513–43.
21. BAER, J. G. (1954). Révision taxonomique et étude biologique des cestodes de la famille des Tetrabothriidae parasites d'oiseaux de haute mer et de mammifères marins. *Mém. Univ. Neuchâtel*. Series in 4to **1**, 1–123.
22. BAILEY, R. E. (1952). The incubation patch of passerine birds. *Condor* **54**, 121–36.
23. BALEN, J. H. VAN (1967). The significance of variations in body weight and wing length of the great tit, *Parus major*. *Ardea* **55**, 1–59.
24. BANG, B. G. (1960). Anatomical evidence for olfactory function in some species of birds. *Nature, Lond.* **188**, 547–9.
25. BANG, B. G. AND COBB, S. (1968). The size of the olfactory bulb in 108 species of birds. *Auk* **85**, 55–61.
26. BANNERMAN, D. A. (1930–51). *The Birds of Tropical West Africa*. Crown Agents, London.
27. BARRETT, J. H. AND HARRIS, M. P. (1965). A count of the gannets on Grassholm in 1964. *Br. Birds* **58**, 201–3.
28. BARRINGTON, E. J. W. (1963). *An Introduction to General and Comparative Endocrinology*. Clarendon Press, Oxford.
29. BARTHOLOMEW, G. A. and DAWSON, W. R. (1952). Body temperature in nestling western gulls. *Condor* **54**, 58–60.
30. BARTHOLOMEW, G. A., LASIEWSKI, R. C. and CRAWFORD, E. C. (1968). Patterns of panting and gular flutter in cormorants, pelicans, owls and doves. *Condor* **70**, 31–4.
31. BEECHER, W. J. (1953). A phylogeny of the Oscines. *Auk* **70**, 270–333.
32. BÉNOIT, J. (1950). Pp. 384–478 in GRASSÉ (ref. 3).
33. BEVEN, G. (1963). Population changes in a Surrey oakwood during fifteen years. *Br. Birds* **56**, 307–23.
34. BEVEN, G. (1964). The food of tawny owls in London. *Lond. Bird Rep.* **29**, 56–72.
35. BEVERLEY, C. T. and HICKLING, R. A. O. (1967). Rooks in Leicestershire. *Rep. Birds Leicestersh. Rutl.* 1965, 4–6.
36. BLAIR, R. H. and TUCKER, B. W. (1941). Nest sanitation. *Br. Birds* **34**, 206–15, 226–35, 250–5.
37. BLONDEL, J. (1967). Réflexions sur les rapports entre prédateurs et proies chez les rapaces. I. Les effets de la prédation sur les populations de proies. *Terre Vie* **5**, 32–62.
38. BOCK, W. J. (1963). The cranial evidence for ratite affinities. *Proc. XIII int. orn. Congr.* **1**, 39–54.
39. BOSSEMA, I. (1968). Recovery of acorns in the European jay *Garrulus g. glandarius* (L.). *Proc. K. ned. Akad. Wet.* C**71**, 1–5.
40. BOURNE, W. R. P. (1955). The birds of the Cape Verde Islands. *Ibis* **97**, 508–56.

41. BRACKHILL, H. (1950). Successive nest sites of individual birds of eight species. *Bird-Banding* **21**, 6–8.
42. BRAESTRUP, F. W. (1966). Social and communal display. *Phil. Trans. R. Soc. Ser. B* **251**, 375–86.
43. BREITENBACH, R. P. and BASKETT, T. S. (1967). Ontogeny of thermoregulation in the mourning dove. *Physiol. Zoöl.* **40**, 207–17.
44. BREMOND, J. C. (1963). Voice and hearing of birds. In *Acoustic Behaviour of Animals*, ed. BUSNEL, R. G. Elsevier, Amsterdam, London and New York.
45. BREWER, R. (1963). Stability in bird populations. *Occ. Pap. C. C. Adams Cent. ecol. Stud.* **7**, 1–12.
46. BRIAN, A. D. (1949). Dominance in the great tit. *Scott. Nat.* **61**, 144–55.
47. BROCKWAY, B. F. (1967). The influence of vocal behaviour on the performer's testicular activity in budgerigars (*Melopsittacus undulatus*). *Wilson Bull.* **79**, 328–34.
48. BROOKS, M. (1947). Interrelations of house wren and Bewick's wren. *Auk* **64**, 624.
49. BROWN, L. H. and WATSON, A. (1964). The golden eagle in relation to its food supply. *Ibis* **106**, 78–100.
50. BROWN, R. G. B. (1967), Species isolation between the herring-gull *Larus argentatus* and lesser black-backed gull *L. fuscus*. *Ibis* **109**, 310–7.
51. BROWN, R. H. J. (1953). The flight of birds. II. Wing function in relation to flight speed. *J. exp. Biol.* **30**, 90–103.
52. BUXTON, E. J. M. (1948). Tits and peanuts. *Br. Birds* **41**, 229–32.
53. CADE, T. J. and GREENWALD, L. (1966). Nasal salt secretion in falconiform birds. *Condor* **68**, 338–50.
54. CALVIN, A. (1960). Olfaction in birds. *Science, N.Y.* **131**, 1263–5.
55. CAMPBELL, B. (1959). Attachment of pied flycatchers *Muscicapa hypoleuca* to nest-sites. *Ibis* **101**, 445–8.
56. CAMPBELL, B. (1968). The Dean nestbox study 1942–1964. *Forestry* **41**, 27–46.
57. CHANCE, B. (1961). Energy-linked cytochrome oxidation in mitochondria. *Nature, Lond.* **189**, 719–25.
58. CHANCE, E. P. (1940). *The Truth about the Cuckoo*. Country Life, London.
59. CHAPIN, J. P. (1954). The calendar of Wideawake Fair. *Auk* **71**, 1–15.
60. CHEW, R. M. (1961). Water metabolism of desert-inhabiting vertebrates. *Biol. Rev.* **36**, 1–31.
61. CHITTY, D. (1967). What regulates bird populations? *Ecology* **48**, 698–701.
62. CHRISTIAN, J. J. and DAVIS, D. E. (1964). Endocrines, behaviour and population. *Science, N.Y.* **146**, 1550–60.
63. CIBA FOUNDATION (1966). *Touch, Heat and Pain*, ed. REUCK, A. V. S. DE and KNIGHT, J., Churchill, London.
64. CLAY, T. (1951). The Mallophaga as an aid to the classification of birds with special reference to the structure of feathers. *Proc. X int. orn. Congr.* 207–15.
65. CLAY, T. (1957). The Mallophaga of birds. *First Symposium on Host Specificity among Parasites of Vertebrates. Neuchâtel.* 120–55.
66. CONDER, P. J. (1956). The territory of the wheatear *Oenanthe oenanthe*. *Ibis* **98**, 453–9.
67. COOMBS, C. F. J. (1961). Rookeries and roosts of the rook and jackdaw in southwest Cornwall. *Bird Study* **8**, 32–7.

68. COTT, H. B. (1968). The status of the Nile crocodile below Murchiso
Falls, 1968. *Bull. int. Un. Conserv. Nat.* **2**, 62–4.

69. COULSON, J. C. (1968). Differences in the quality of the birds nestin
in the centre and on the edges of a colony. *Nature, Lond.* **217**
478–9.

70. COULSON, J. C. and WHITE, E. (1956). A study of colonies of the kittiwak
Rissa tridactyla (L.). *Ibis* **98**, 63–79.

71. CREED, E. R., DOWDESWELL, W. H., FORD, E. B. and MCWHIRTER, K. G. M
(1962). Evolutionary studies on *Maniola jurtina*: the English mainland. *Heredity, Lond.* **17**, 237–65.

72. DAVIS, D. E. (1946). A seasonal analysis of mixed flocks of birds in Brazi
Ecology **27**, 168–81.

73. DAVIS, P. (1965). Recoveries of swallows ringed in Britain and Irelanc
Bird Study **12**, 151–69.

74. DAVIS, T. A. W. (1953). An outline of the ecology and breeding seasor
of birds of the lowland forest region of British Guiana. *Ibis* **9**
450–67.

75. DAVIS, W. B. (1933). The span of the nesting season of birds in Butt
County, California, in relation to their food. *Condor* **35**, 151–4.

76. DARWIN, C. R. (1842). In *Evolution by Natural Selection*, (1958). Can
bridge University Press, Cambridge.

77. DAWSON, W. R. (1958). Relation of oxygen consumption and evaporati
water loss to temperature in the cardinal. *Physiol. Zoöl.* **31**, 37–48.

78. DAWSON, W. R. and EVANS, F. C. (1960). Relation of growth and develo
ment to temperature regulation in nestling vesper sparrows. *Cond*
62, 329–40.

79. DAWSON, W. R., SHOEMAKER, V. H., TORDOFF, H. B. and BORUT, A. (196
Observations on the metabolism of sodium chloride in the red cros
bill. *Auk* **82**, 606–23.

80. DE BEER, G. R. (1954). *Archaeopteryx lithographica*. A study based on t
British Museum specimen. British Museum, London.

81. DE BEER, G. R. (1956). Evolution of the Ratites. *Bull. Br. Mus. nat. Hi*
(*Ser. D*) **4**, 57–70.

82. DE BRACY, L. (1946). Auprès du nid de l'hirondelle de chemin
Hirundo rustica rustica Linné. *Gerfaut* **36**, 133–93.

83. DE KOCK, L. L. (1959). The carotid body system of the higher vertebrat
Acta anat. **37**, 265–79.

84. DISNEY, H. J. DE S., LOFTS, B. and MARSHALL, A. J. (1959). Duration of t
regeneration period of the internal reproductive rhythm in a xer
philous equatorial bird, *Quelea quelea*. *Nature, Lond.* **184**, 1659–60

85. DOBBS, A. (1964). Rook numbers in Nottinghamshire over 35 years. *B*
Birds **57**, 360–4.

86. DONAHUE, M., KÄLLANDER, H., MAWDSLEY, T. and SVENSSON, S. (196
Ornithological observations in Finmark in the summers of 1962 a
1964. *Sterna* **7**, 121–32.

87. DONNER, K. O. (1951). The visual acuity of some passerine birds. *Ac*
zool. fenn. **66**, 1–40.

88. DOWSETT, R. J. (1968). Migrants at malamfatori, Lake Chad, spri
1968. *Bull. niger. orn. Soc.* **5**, 53–6.

89. DUKES, H. H. (1955). *The Physiology of Domestic Animals*, 7th ec
Baillière, Tindall and Cox, London.

90. DURANGO, S. (1953). Om fåglarnas val av häckningsbiotoper. *Sve*
faun. Revy **15**, 58–69.

91. DURANGO, S. (1956). Territory in the red-backed shrike *Lanius collurio*. *Ibis* **98**, 476–84.
92. DYE, J. A. (1955). In DUKES (ref. 89).
93. EASTWOOD, E. (1967). *Radar Ornithology*. Methuen, London.
94. EDWARDS, V. C. WYNNE- (1962). *Animal Dispersion in Relation to Social Behaviour*. Oliver and Boyd, Edinburgh.
95. EDWARDS, V. C. WYNNE-, LOCKLEY, R. M., and SALMON, H. M. (1936). The distribution and numbers of breeding gannets (*Sula bassana*). *Br. Birds* **29**, 162–76.
96. ELIASSEN, E. (1963). Preliminary results from new methods of investigating the physiology of birds during flight. *Ibis* **105**, 234–7.
97. ELTON, C., DAVIS, D. H. S., and FINDLAY, G. M. (1935). An epidemic among voles (*Microtus agrestis*) on the Scottish border in the spring of 1934. *J. Anim. Ecol.* **4**, 277–85.
98. EMLEN, J. T. (1941). An experimental analysis of the breeding cycle of the tricolored redwing. *Condor* **43**, 209–19.
99. EMLEN, J. T. and PENNEY, R. L. (1964). Distance navigation in the Adélie penguin. *Ibis* **106**, 417–31.
100. EMLEN, S. T. (1967). Migratory orientation in the indigo bunting, *Passerina cyanea*. I. Evidence for use of celestial cues. *Auk* **84**, 309–42.
101. ENGLISH, T. M. S. (1923). On the greater length of day in high latitudes as a reason for spring migration. *Ibis* **65**, 418–23.
102. EVANS, P. R. (1968). Autumn movements and orientation of waders in northeast England and southern Scotland studied by radar. *Bird Study* **15**, 53–64.
103. EVANS, P. R. (1968). Re-orientation of small passerine night-migrants after displacement: evidence from Kramer-type cages and ringing recoveries. *Ibis* **110**, 412.
104. EVERETT, S. D. (1966). In HORTON-SMITH and AMOROSO (ref. 325).
105. FALCONER, D. S. (1964). *Quantitative Genetics*, Oliver and Boyd, Edinburgh.
106. FALLA, R. A. (1936). Arctic birds as migrants in New Zealand. *Rec. Auckland Inst. Mus.* **2** (1), 3–14.
107. FALLA, R. A. (1937). Birds. *Rep. B.A.N.Z. antarct. Res. Exped. 1929–31* B2.
108. FARNER, D. S. (1959). Photoperiodic control of annual gonadal cycles. In *Photoperiodism and Related Phenomena in Plants and Animals*, ed. WITHROW R. B., American Association for the Advancement of Science. Bailey Bros., London.
109. FARNER, D. S. (1967). The control of avian reproductive cycles. *Proc. XIV int. orn. Congr.* 107–33.
110. FICKEN, M. S. and FICKEN, R. W. (1967). Age-specific differences in the breeding behavior and ecology of the American redstart. *Wilson Bull.* **79**, 188–99.
111. FISHER, H. I. (1966). Airplane–albatross collisions on Midway Atoll. *Condor* **68**, 229–42.
112. FISHER, J. (1966). The fulmar population of Britain and Ireland, 1959. *Bird Study* **13**, 5–76.
113. FISHER, J. and VEVERS, H. G. (1943–44). The breeding distribution, history and population of the north Atlantic gannet (*Sula bassana*). *J. Anim. Ecol.* **12**, 173–213, and **13**, 49–62.
114. FOX, H. M. and VEVERS, G. (1960). *The Nature of Animal Colours*. Sidgwick and Jackson, London.

115. FOX, S. and MORRIS, T. R. (1958). Flash lighting for egg production. *Nature, Lond.* **182**, 1752–3.
116. FRIEDMANN, H. and KERN, J. (1956). The problem of cerophagy or wax-eating in the honey-guides. *Q. Rev. Biol.* **31**, 19–30.
117. FRITH, H. J. (1956). Temperature regulation in the nesting mounds of the Mallee-fowl, *Leiopa ocellata* Gould. *C.S.I.R.O. Wildl. Res.* **1**, 79–95.
118. FROMME, H. G. (1961). Untersuchungen über das Orientierungs-vermögen nächtlich ziehender Kleinvögel (*Erithacus rubecula, Sylvia communis*). *Z. Tierpsychol.* **18**, 205–20.
119. FRY, C. H. (1969). The recognition and treatment of venomous and non-venomous insects by small bee-eaters. *Ibis* **111**, 23–9.
120. GARROD, A. H. (1874). On certain muscles of the thigh in birds and on their use in classification. II. *Proc. zool. Soc. Lond.* 111–23.
121. GAUSE, C. F. (1934). *The Struggle for Existence*. Williams and Wilkins Baltimore.
122. GAUSE, G. F. (1935). *Vérifications expérimentales de la théorie mathematique de la lutte pour la vie*. Actualités scientifiques et industrielles. Hermann et Cie, Paris.
123. GEORGE, J. C. and JYOTI, D. (1958). Studies on the structure and physiology of the flight muscles of birds. II. The relative reduction of fat and glycogen in the pectoralis major muscle during sustained activity *J. Anim. Morph. Physiol.* **5**, 57–60.
124. GEORGE, J. C. and NAIK, R. M. (1958). Relative distribution and chemical nature of the fuel store of the two types of fibres in the pectoralis major muscle of the pigeon. *Nature, Lond.* **181**, 709–10.
125. GÉROUDET, P. (1966). Premières consequences ornithologiques de l'introduction de la moule zebrée *Dreissena polymorpha* dans le lac Léman. *Nos Oiseaux* **28**, 301–7.
126. GIBB, J. A. (1950). The breeding biology of the great and blue titmice *Ibis* **92**, 507–39.
127. GIBB, J. A. (1956). Territory in the genus *Parus*. *Ibis* **98**, 420–9.
128. GIBB, J. A. (1958). Predation by tits and squirrels on the eucosmid *Ernarmonia conicolana* (Heyl.). *J. Anim. Ecol.* **27**, 375–96.
129. GIBB, J. A. (1966). Tit predation and the abundance of *Ernarmonia conicolana* (Heyl.) on Weeting Heath, Norfolk, 1962–63. *J. Anim. Ecol.* **35**, 43–53.
130. GILBERT, P. W. and BOND, C. F. (1949). Comparative study of haemoglobin and red cells of representative diving and dabbling ducks *Science, N.Y.* **109**, 36–7.
131. GLOVER, R. S., COOPER, G. A. and BROWN, W. W. (1956). An ecological survey of the drift-net herring fishery off the north-east coast of Scotland. I: Sampling by the commercial fishing fleet. *Bull. mar. Ecol.* **4**, 141–78.
132. GOODACRE, M. J. (1959). The origin of winter visitors to the British Isles. *Bird Study* **6**, 103–8 (chaffinch) and 180–92 (starling).
133. GOODWIN, D. (1956). Further observations on the behaviour of the jay *Garrulus glandarius*. *Ibis* **98**, 186–219.
134. GRABER, R. R. and GRABER, J. W. (1963). A comparative study of bird populations in Illinois, 1906–1909 and 1956–1958. *Bull. Ill. St. nat. Hist. Surv.* **28**, 383–526.
135. GRAHAM, R. R. (1932). The part played by the emarginated primaries and the alula in the flight of birds. *Bull. Br. Orn. Club* **52**, 68–79.

136. GRAY, J. A. B. (1963). Coding in systems of primary receptor neurons. *Symp. Soc. exp. Biol.* **16**, 345–54.
137. GRIFFIN, D. R. (1953). Accoustic orientation in the oilbird, *Steatornis. Proc. natn. Acad. Sci. U.S.A.* **39**, 884–93.
138. GRIFFIN, D. R. and HOCK, R. J. (1949). Airplane observations of homing birds. *Ecology* **30**, 176–98.
139. GRIFFITHS, M. NORTON- (1967). Some ecological aspects of the feeding behaviour of the oyster-catcher *Haematopus ostralagus* on the edible mussel *Mytilus edulis*. *Ibis* **109**, 412–24.
140. GRINNELL, J. (1904). The origin and distribution of the chestnut-backed chickadee. *Auk* **21**, 364–82.
141. GUDMUNDSSON, F. (1952). The effects of the recent climatic changes on the bird life of Iceland. *Proc. X int. orn. Congr.* 502–14.
142. HAARTMAN, L. VON (1954). Clutch size in polygamous species. *Acta XI Congr. int. orn.* 450–3.
143. HAARTMAN, L. VON (1956). Territory in the pied flycatcher *Muscicapa hypoleuca*. *Ibis* **98**, 460–75.
144. HAARTMAN, L. VON (1956). Der Einfluss der Temperatur auf den Brut-rhythmus experimentall nachgewiesen. *Ornis fenn.* **33**, 100–7.
145. HAARTMAN, L. VON (1958). The incubation rhythm of the female pied flycatcher (*Ficedula hypoleuca*) in the presence of the male. *Ornis fenn.* **35**, 71–6.
146. HAARTMAN, L. VON (1958). The decrease of the corncrake (*Crex crex*). *Commentat. biol.* **18**, 1–29.
147. HAARTMAN, L. VON (1960). The *ortstreue* of the pied flycatcher. *Proc. XII int. orn. Congr.* **1**, 266–73.
148. HAARTMAN, L. VON (1967). Geographical variation in the clutch-size of the pied flycatcher. *Ornis fenn.* **44**, 89–98.
149. HAFTORN, S. (1954–56). Contribution to the food biology of the tits especially about storing of surplus food. I. The crested tit (*Parus c. cristatus L.*) *K. norske Vidensk. Selsk. Skr.* 1953, 1–123. II. The coal tit (*Parus a. ater L.*). Ibid. 1956 (2) 1–52. III. The willow tit (*Parus atricapillus L.*). Ibid. 1956 (3) 1–80. IV. A comparative analysis of *Parus atricapillus L., P. cristatus L.* and *P. ater L.* Ibid. 1956 (4) 1–54.
150. HAMMOND, J. (1941). Fertility in mammals and birds. *Biol. Rev*, **16**, 177–90.
151. HANSEN, E. W. (1966). Squab induced crop growth in ring dove foster parents. *J. comp. physiol Psychol.* **62**, 120–2.
152. HARPER, W. G. (1957). 'Angels' on centimetric radar caused by birds. *Nature, Lond.* **180**, 847–9.
153. HARRIS, M. P. (1964). Aspects of the breeding biology of the gulls *Larus argentatus, L. fuscus* and *L. marinus*. *Ibis.* **106**, 432–56.
153a. HARTLEY, P. H. T. (1947). The food of the long-eared owl in Iraq. *Ibis* **89**, 566–69.
154. HARTSHORNE, C. (1958). Some biological principles applicable to song-behaviour. *Wilson Bull.* **70**, 41–56.
155. HASLEWOOD, G. A. D. (1964). The biological significance of chemical differences in bile salts. *Biol. Rev.* **39**, 537–74.
156. HAZELHOFF, E. H. (1951). Structure and function of the lung of birds. *Poult. Sci.* **30**, 3–10. (Translation of *Versl. gewone Vergad. Afd. K. ned. Akad. Wet.* **53**, 391–400 (1943)).
157. HEALD, P. J. (1966). In HORTON-SMITH and AMOROSO (ref. 325).
158. HEILMAN, G. (1926). *The Origin of Birds*. Witherby, London.

159. HENDERSON, M. (1968). The rook population of a part of west Cheshire 1944–1968. *Bird Study* **15**, 206–8.
160. HENSLEY, M. M. and COPE, J. B. (1951). Further data on removal and repopulation of the breeding birds in a spruce-fir forest. *Auk* **68**, 483–93.
161. HILDÉN, O. (1965). Habitat selection in birds. *Annls zool. fenn.* **2**, 53–75.
162. HINDE, R. A. (1958). The nest-building behaviour of domesticated canaries. *Proc. zool. Soc. Lond.* **131**, 1–48.
163. HINDE, R. A. and FISHER, J. (1949). The opening of milk bottles by birds. *Br. Birds* **42**, 347–57.
164. HOFFMAN, K. (1958). Repetition of an experiment on bird orientation. *Nature, Lond.* **181**, 1435–7.
165. HOHN, E. O. (1962). A possible endocrine basis of brood parasitism. *Ibis* **104**, 418–21.
166. HOLMGREN, N. (1955). Studies on the phylogeny of birds. *Acta zool. Stockh.* **36**, 243–328.
167. HOWARD, H. E. (1907–1914). *The British Warblers: a history with problems of their lives.* Porter, London.
168. HOWARD, H. E. (1920). *Territory in Bird Life.* Murray, London.
169. HOWELL, T. R. and BARTHOLOMEW, G. A. (1959). Further experiments on torpidity in the poorwill. *Condor* **61**, 180–5.
170. HOWELL, T. R. and BARTHOLOMEW, G. A. (1962). Temperature regulation in the sooty tern *Sterna fuscata. Ibis* **104**, 98–105.
171. HUSTON, T. M. and NELBANDOV, A. V. (1953). Neurohumoral control of the pituitary in the fowl. *Endocrinology* **52**, 149–56.
171a. IMMELMANN, K. (1963). Drought adaptations in Australian desert birds. *Proc. XIII int. orn. Congr.* **2**, 649–57.
172. JENKINS, D., WATSON, A. and MILLER, G. R. (1964). Predation and red grouse populations. *J. appl. Ecol.* **1**, 183–95.
173. JENKINS, D., WATSON, A. and PICOZZI, N. (1963). Red grouse chick survival in captivity and in the wild. *Trans. VI Congr. int. Union Game Biologists* 63–70.
174. JEUNIAUX, C. (1962). Digestion de la chitine chez les oiseaux et les mammifères. *Annls. Soc. r. zool. Belg.* **92**, 27–45.
175. JOHNSTON, R. F. (1956). Population structure in salt-marsh song-sparrows. *Condor* **58**, 24–44; 254–72.
176. JONES, R. E. and LEOPOLD, A. S. (1967). Nesting interference in a dense population of wood ducks. *J. Wildl. Mgmt.* **31**, 221–8.
177. KALISCHER, O. (1905). Das Grosshirn der Papageien in anatom. und physiol. Beziehung. *Abh. preuss. Akad. Wiss. Anhang. Abt.* **4**, 1–105.
178. KEAST, J. A. and MARSHALL, A. J. (1954). Reproduction in Australian desert birds. *Proc. zool. Soc. Lond.* **124**, 493–9.
179. KEILIN, JOAN (1959). The biological significance of uric acid and guanine excretion. *Biol. Rev.* **34**, 265–96.
180. KENDEIGH, S. C. (1944). Effect of air temperature on the rate of energy metabolism in the English sparrow. *J. exp. Zool.* 96, 1–16.
181. KENDEIGH, S. C. and BALDWIN, S. P. (1928). Development of temperature control in nestling house wrens. *Am. Nat.* **62**, 249–78.
182. KENDEIGH, S. C., WEST, G. C. and COX, G. W. (1960). Annual stimulus for migration in birds. *Anim. Behav.* **8**, 180–5.
183. KENT, A. K. (1964). The breeding habitats of the reed bunting and yellowhammer in Nottinghamshire. *Bird Study* **11**, 123–7.

184. KING, A. S. and PAYNE, D. C. (1962). The maximum capacities of the lungs and air-sacs of *Gallus domesticus*. *J. Anat.* **96**, 495–503.
185. KING, J. R. (1968). Cycles of fat deposition and molt in white-crowned sparrows in constant environmental conditions. *Comp. Biochem. Physiol.* **24**, 827–37.
186. KING, J. R. and FARNER, D. S. (1959). Premigratory changes in body weight and fat in wild and captive male white-crowned sparrows. *Condor* **61**, 315–24.
187. KING, J. R. and FARNER, D. S. (1965). Studies of fat deposition in migratory birds. *Ann. N.Y. Acad. Sci.* **131**, 422–40.
188. KING, J. R. FOLLETT, B. K., FARNER, D. S. and MORTON, M. C. (1966). Annual gonadal cycles and pituitary gonadotropins in *Zonotrichia leucophrys gambellii*. *Condor* **68**, 476–87.
189. KLUYVER, H. N. (1963). The determination of reproductive rates in Paridae. *Proc. XIII int. orn. Congr.* **2**, 706–16.
190. KLUYVER, H. N. and TINBERGEN, L. (1953). Territory and the regulation of density in titmice. *Archs néerl. Zool.* **10**, 265–89.
191. KONISHI, M. (1965). The role of auditory feedback in the control of vocalization in the white-crowned sparrow. *Z. Tierpsychol.* **22**, 770–83.
192. KOPISCHKE, E. D. (1966). Selection of calcium- and magnesium-bearing grit by pheasants in Minnesota. *J. Wildl. Mgmt* **30**, 276–9.
193. KRAMER, G. (1951). Eine neue Methode zur Erforschung der Zugorientierung und die bisher damit erzielten Ergebnisse. *Proc. X int. orn. Congr.* 269–80.
194. LACK, D. (1946). *The Life of the Robin*, 2nd edn. Witherby, London.
195. LACK, D. (1948). Natural selection and family size in the starling. *Evolution, Lancaster, Pa.* **2**, 95–110.
196. LACK, D. (1954). *The Natural Regulation of Animal Numbers*. Clarendon Press, Oxford.
197. LACK, D. (1964). A long-term study of the great tit (*Parus major*). *J. Anim. Ecol.* **33**, suppl., 159–73.
198. LACK, D. and SOUTHERN H. N. (1949). Birds on Teneriffe. *Ibis* **91**, 607–26.
199. LAPICQUE, L. (1909). Le poids de l'encéphale dans les différents groupes d'oiseaux. *Bull. Mus. natn. Hist. nat., Paris.* **15**, 408–12.
200. LASIEWSKI, R. C. (1961). Energy metabolism of humming-birds. *Am. Zool.* **1**, 368.
201. LASIEWSKI, R. C. (1963). Oxygen consumption of torpid, resting, active and flying humming-birds. *Physiol. Zoöl.* **36**, 122–40.
202. LASIEWSKI, R. C. (1964). Body temperature, heart and breathing rate, and evaporative water loss in humming-birds. *Physiol. Zoöl.* **37**, 212–23.
203. LASIEWSKI, R. C. and DAWSON, W. R. (1964). Physiological responses to temperature in the common nighthawk. *Condor* **66**, 477–90.
204. LASIEWSKI, R. C. and DAWSON, W. R. (1967). A re-examination of the relation between standard metabolic rate and body weight in birds. *Condor* **69**, 13–23.
205. LASIEWSKI, R. C., HUBBARD, S. H. and MOBERLY, W. R. (1964). Energetic relationships of a very small passerine bird. *Condor* **66**, 212–20.
206. LASIEWSKI, R. C. and LASIEWSKI, R. J. (1967). Physiological responses of blue-throated and Rivoli's humming-birds. *Auk* **84**, 34–48.
207. LEOPOLD, A. S. (1953). Intestinal morphology of gallinaceous birds in relation to food habits. *J. Wildl. Mgmt.* **17**, 197–203.

208. LEWTY, R. A. CARR- (1946). Aviation and ornithology in Lakeland. *Trans. Carlisle nat. Hist. Soc.* **7**, 1–28.

209. LINCOLN, F. C. (1939). *The Migration of American Birds.* Doubleday, New York.

210. LIND, H. (1955). Bidrag til Solsortens (*Turdus m. merula* (L.)) biologi. *Dansk orn. Foren. Tidsskr.* **49**, 76–113.

211. LINSDALE, J. M. (1933). The nesting season of birds in Doniphan County, Kansas. *Condor* **35**, 155–60.

212. LLOYD, H. G. (1968). Observations on nut-selection by a hand-reared grey squirrel (*Sciurus carolinensis*). *J. Zool.* **155**, 240–4.

213. LOFTS, B. (1964). Evidence of an autonomous reproductive rhythm in an equatorial bird (*Quelea quelea*). *Nature, Lond.* **201**, 523–4.

214. LOFTS, B. and MARSHALL, A. J. (1960). The experimental regulation of *zugunruhe* and the sexual cycle in the brambling *Fringilla montifringilla. Ibis* **102**, 209–14.

215. LOFTS, B., MARSHALL, A. J., and WOLFSON, A. (1963). The experimental demonstration of pre-migration activity in the absence of fat deposition in birds. *Ibis.* **105**, 99–105.

216. LOFTS, B. and MURTON, R. K. (1968). Photoperiodic and physiological adaptations regulating avian breeding cycles and their ecological significance. *J. Zool.* **155**, 327–94.

217. LOFTS, B., MURTON, R. K. and WESTWOOD, N. J. (1967). Photoresponses of the wood-pigeon *Columba palumbus* in relation to the breeding season. *Ibis* **109**, 338–51.

218. LOMAS, P. D. R. (1968). The decline of the rook population of Derbyshire. *Bird Study* **15**, 198–205.

219. LORENZ, F. W. (1966). Behaviour of spermatozoa in the oviduct in relation to fertility. In HORTON-SMITH and AMOROSO (ref. 325).

220. LOWE, P. R. (1935). On the relationships of the Struthiones to the dinosaurs and to the avian class, with special reference to the position of *Archaeopteryx. Ibis* **77**, 398–432.

221. MCCLURE, H. E. (1967). The composition of mixed species flocks in lowland and sub-montane forest in Malaya. *Wilson Bull.* **79**, 131–54.

222. MCDONALD, D. I. (1968). Bird orientation: a method of study. *Science, N.Y.* **161**, 486–7.

223. MACDONALD, J. D. (1959). Old and new lines in taxonomy. *Ibis* **101**, 330–4.

224. MCFARLAND, L. Z. and GEORGE, H. (1966). Preference of selected grains by geese. *J. Wildl. Mgmt,* **30**, 9–13.

225. MCNULTY, F. (1967). *The Whooping Crane.* Longmans, Green, London.

226. MAGEE, J. D. (1965). The breeding distribution of the stonechat in Britain and the causes of its decline. *Bird Study* **12**, 83–9.

227. MANWELL, C. (1960). Comparative physiology: blood pigments. *A. Rev. Physiol.* **22**, 191–244.

228. MANWELL, C., BAKER, C. M. A., ROSLOWSKY, J. D. and FOGHT, D. (1963). Molecular genetics of avian proteins. II. Control genes and structural genes for embryonic and adult hemoglobins. *Proc. natn. Acad. Sci. U.S.A.* **49**, 496–503.

229. MARCHANT, S. (1959). The breeding season in S.W. Ecuador. *Ibis.* **101**, 137–52.

230. MARLER, P. (1956). Territory and individual distance in the chaffinch *Fringilla coelebs. Ibis* **98**, 496–501.

231. MARLER, P. (1957). Specific distinctiveness in the communication signals of birds. *Behaviour* **11**, 13–39.
232. MARLER, P. (1967). Comparative study of song development in sparrows. *Proc. XIV int. orn. Congr.* 231–44.
233. MARSHALL, A. J. (1950). The function of the bower of the satin bower-bird in the light of experimental modifications of the breeding cycle. *Nature, Lond.* **165**, 388–90.
234. MARSHALL, A. J. (1952). Non-breeding among arctic birds. *Ibis* **94**, 310–33.
235. MARSHALL, A. J. (1954). *Bower-Birds.* Clarendon Press, Oxford.
235a. MARSHALL, A. J. (1959). Internal and environmental control of breeding. *Ibis.* **101**, 456–78.
236. MARSHALL, A. J. and COOMBS, C. F. J. (1957). The interaction of environmental, internal and behavioral factors in the rook *Corvus frugilegus* Linnaeus. *Proc. zool. Soc. Lond.* **128**, 545–89.
237. MARSHALL, A. J. and DISNEY, H. J. DE S. (1959). Experimental induction of the breeding cycle in a xerophilous bird. *Nature, Lond.* **180**, 647–9.
238. MARSHALL, A. J. and SERVENTY, D. L. (1959). Experimental demonstration of an internal rhythm of reproduction in a trans-equatorial migrant (the short-tailed shearwater *Puffinus tenuirostris*). *Nature, Lond.*, **184**, 1704–5.
239. MARVIN, P. H. (1964). Birds on the rise. *Bull. ent. Soc. Am.* **10**, 194–6.
240. MATTHEWS, G. V. T. (1955). *Bird Navigation.* Cambridge University Press, Cambridge.
241. MATTHEWS, G. V. T. (1963). The astronomical basis of 'nonsense' orientation. *Proc. XIII int. orn. Congr.* **1**, 415–29.
241a. MAY, D. J. (1949). Studies on a community of willow-warblers. *Ibis* **91**, 25–54.
242. MAYFIELD, H. (1962). 1961 decennial census of the Kirtland's warbler. *Auk* **79**, 173–82.
243. MAYR, E. (1942). *Systematics and the Origin of Species.* Columbia University Press, New York.
244. MEINERTZHAGEN, R. (1955). The speed and altitude of bird flight (with notes on other animals). *Ibis* **97**, 81–117.
245. MEISE, W. (1928). Die Verbreitung der Aaskrahe (Formenkreis *Corvus corone* L.). *J. Orn., Lpz.* **76**, 1–203.
246. MEISE, W. (1963). Verhalten der Straussartigen Vögel und Monophylie der Ratitae. *Proc. XIII int. orn. Congr.* **1**, 115–25.
247. MEITES, J. and NICOLL, C. S. (1966). Adenohypophysis: prolactin. *A. Rev. Physiol.* **28**, 57–88.
248. MERREM, B. (1816). Tentamen systematis naturalis avium. *Abh. preuss. Akad. Wiss.* 1812–13, 237–59.
249. MERTENS, J. A. L. (1969). The influence of brood size on the energy metabolism and water loss of nestling great tits *Parus major major.* *Ibis* **111**, 11–6.
250. MICHELSEN, W. J. (1959). Procedure for studying olfactory discrimination in pigeons. *Science, N.Y.* **130**, 630–1.
251. MICHENER, M. C. and WALCOTT, C. (1966). Navigation of single homing pigeons: airplane observations by radio tracking. *Science, N.Y.* **154**, 410–3.
252. MILLIKAN, G. C. and BOWMAN, R. I. (1967). Observations on Galápagos tool-using finches in captivity. *Living Bird* **6**, 23–41.

253. MOFFAT, C. B. (1903). The spring rivalry of birds. *Ir. Nat.* **12**, 152–66.
254. MOORE, N. W. (1957). The past and present status of the buzzard in the British Isles. *Br. Birds* **50**, 173–97.
255. MOREAU, R. E. (1930). On the age of some races of birds. *Ibis* **72**, 229–39.
256. MOREAU, R. E. (1952). The place of Africa in the palaearctic migration system. *J. Anim. Ecol.* **21**, 250–71.
257. MOREL, M. Y. (1967). Les oiseaux tropicaux élèvent-ils autant de jeunes qu'ils peuvent en nourir? Le cas de *Lagonostricta senegala. Terre Vie* **114**, 77–82.
258. MORGAN, C. L. (1896). *Habit and Instinct.* Edward Arnold, London.
259. MORLEY, AVRIL (1941). The behaviour of a group of resident starlings (*Sturnus v. vulgaris Linn.*) from October to March. *Naturalist, Hull* **66**, 55–61.
260. MORSE, D. H. (1968). The use of tools by brown-headed nuthatches. *Wilson Bull.* **80**, 220–4.
261. MOSS, R. (1969). A comparison of red grouse (*Lagopus l. scoticus*) stocks with the production and nutritive value of heather (*Calluna vulgaris*). *J. Anim. Ecol.* **38**, 103–22.
262. MOUNTFORD, M. D. (1968). The significance of litter-size. *J. Anim. Ecol.* **37**, 363–7.
263. MOUNTFORT, R. (1956). The territorial behaviour of the hawfinch *Coccothraustes coccothraustes. Ibis* **98**, 490–5.
264. MÜLLER, B. (1908). The air-sacs of the pigeon. *Smithson. misc. Collns* **50**, 365–414.
265. MURIE, J. (1874). On the nature of the sacs vomited by hornbills. *Proc. zool. Soc. Lond.* 420–5.
266. MURTON, R. K. (1958). The breeding of wood-pigeon populations. *Bird Study* **5**, 157–83.
267. MURTON, R. K. (1966). Natural selection and the breeding seasons of the stock-dove and wood-pigeon. *Bird Study* **13**, 311–27.
268. MURTON, R. K., ISAACSON, A. J. and WESTWOOD, N. J. (1966). The relationships between wood-pigeons and their clover food supply and the mechanism of population control. *J. appl. Ecol.* **3**, 55–96.
269. NALBANDOV, A. V., HOCHHAUSER, M. and DUGAS, M. (1945). A study of the effect of prolactin on broodiness and on cock testes. *Endocrinology* 36, 251–8.
270. NELSON, D. M. and NALBANDOV, A. V. (1966). Hormone control of ovulation. In HORTON-SMITH and AMOROSO (ref. 325).
271. NELSON, J. B. (1964). Factors influencing clutch-size and chick growth in the north Atlantic gannet *Sula bassana. Ibis* **106**, 63–77.
272. NEWTON, I. (1964). Bud-eating by bullfinches in relation to the natural food-supply. *J. appl. Ecol.* **1**, 265–79.
273. NICHOLSON, E. M. (1938). The index of heron population 1938. *Br. Birds* **32**, 138–44.
274. NICOLAI, J. (1964). Der Brutparasitismus der Viduinae als ethologisches Problem: Prägungsphänomene als Faktoren der Rassen- und Artbildung. *Z. Tierpsychol.* **21**, 129–204.
274a. NORDSTRÖM, G. (1953). Boniteringsundersökningar över häckfågelfaunan inom tre olika skogsområden under en följd av fem somrar. *Ornis fenn.* **30**, 56–67.
275. NORRIS, C. A. (1960). The breeding distribution of thirty bird species in 1952. *Bird Study* **7**, 129–84.

276. NOTTEBOHM, F. (1968). Auditory experience and song development in the chaffinch *Fringilla coelebs*. *Ibis* **110**, 549–68.

277. NOVAKOWSKI, N. S. (1966). Whooping crane population dynamics on the nesting grounds, Wood Buffalo National Park, Northwest Territories, Canada. *Can. Wildl. Service. Rept.* Ser. **1**.

278. ODUM, E. P. (1945). The heart-rate of small birds. *Science, N.Y.* **101**, 153–4.

279. ODUM, E. P. and CONNELL, C. E. (1956). Lipid levels in migrating birds. *Science, N.Y.* **123**, 892–4.

280. OLNEY, P. S. J. (1965). The food and feeding habits of shelduck *Tadorna tadorna*. *Ibis* **107**, 527–32.

281. OWEN, R. (1863). On the *Archaeopteryx* of von Mayer, with a description of the fossil remains of a long-tailed species from the lithographic stone of Solenhaufen. *Phil. Trans. R. Soc.* **153**, 33–47.

282. OWRE, O. T. and NORTHINGTON, P. O. (1961). Indication of the sense of smell in the turkey vulture, *Cathartes aura* (L.), from feeding tests. *Am. Midl. Nat.* **66**, 200–5.

283. PAYNE, R. B. (1968). Mimicry and relationships in the indigobirds or combassous of Nigeria. *Bull. Niger. orn. Soc.* **5**, 57–60.

284. PEAL, R. E. F. (1968). The distribution of the wryneck in the British Isles 1964–66. *Bird Study* **15**, 111–26.

285. PEIPONEN, V. A. (1957). Wechselt der Birkenzeisig, *Carduelis flammea* (L.), sein Brutgebiet während des Sommers.? *Ornis. fenn.* **34**, 41–64.

286. PEMBREY, M. S., GORDON, M. H. and WARREN, R. (1894–95). On the response of the chick, before and after hatching, to changes of external temperature. *J. Physiol., Lond.* **17**, 331–48.

287. PERRINS, C. M. (1965). Population fluctuations and clutch-size in the great tit, *Parus major*. *J. Anim. Ecol.* **34**, 601–47.

288. PESTILLINI, G. (1941). Influenza della temperatura dell'acqua sull'apnea da immersione nell'anatra. *Archo Fisiol.* **41**, 411–24.

289. PINTO, O. (1953). Sobre a coleção Carlos Estevo de peles, ninhos e ovos das aves de Belém (Pará). *Papéis Dep. Zool. S Paulo* **11**, 111–222.

290. PORTMAN, A. (1946–47). Etudes sur la cérébralisation chez les oiseaux. *Alauda* **14**, 2–20; **15**, 111–222.

291. PRESTT, I. and MILLS, D. H. (1966). A census of the great crested grebe in Britain 1965. *Bird Study* **13**, 163–203.

292. PUMPHREY, R. J. (1948). The sense organs of birds. *Ibis* **90**, 171–99.

293. PYCRAFT, W. P. (1898). A contribution towards our knowledge of the morphology of the owls. *Trans. Linn. Soc. Lond.* **7**, 223–75.

294. QUILLIAM, T. A. (1966). Unit design and array patterns in receptor organs. In CIBA FOUNDATION (ref. 63).

295. RANDALL, W. C. (1943). Factors influencing the temperature regulation of birds. *Am. J. Physiol.* **139**, 56–63.

296. RATCLIFFE, D. A. (1963). The status of the peregrine in Great Britain. *Bird Study* **10**, 56–90.

297. RATCLIFFE, D. A. (1967). The peregrine situation in Great Britain 1965–66. *Bird Study* **14**, 238–46.

298. REED, C. I. and REED, B. P. (1928). The mechanism of pellet formation in the great horned owl (*Bubo virginianus*). *Science, N.Y.* **68**, 359–60.

299. RETZIUS, G. (1884). Das Gehörorgan der Wirbelthiere. II. Das Gehörorgan der Reptilien, der Vögel und die Säugethiere. Samson & Wallin, Stockholm.

300. RICKLEFS, R. E. and HAINSWORTH, F. R. (1968). Temperature regulation

in nestling cactus wrens: the development of homeothermy. *Condor* **70**, 121–7.

301. ROBERTSON, W. B. (1965). Migrations of sooty terns. *Niger. Fld* **30**, 190–1.
302. ROBSON, F. D. (1947). Kiwis in captivity as told to Robert Gibbings. *Hawkes' Bay Art Gallery Mus. Publ.* No. 1. Napier, N.Z.
303. ROMANES, G. J. (1883). *Mental Evolution in Animals*, Kegan Paul, Trench, London.
304. ROMANOFF, A. L. and ROMANOFF, A. J. (1967). *Biochemistry of the Avian Embryo.* John Wiley, New York and London.
305. ROWAN, M. K. (1966). Territory as a density-regulating mechanism in some South African birds. *Ostrich Suppl.* **6**, 397–408.
306. ROWAN, W. (1938). Light and seasonal reproduction in animals. *Biol. Rev.* **13**, 374–402.
307. SAGE. B. L. and NAN, B. S. (1963). The population ecology of the rook in Hertfordshire. *Trans. Herts. nat. Hist. Soc. Fld Club* **25**, 226–44.
308. SALT, G. W. (1964). Respiratory evaporation in birds. *Biol. Rev.* **39**, 112–36.
309. SCHMIDT-KOENIG, K. (1964). Initial orientation and distance of displacement in pigeon homing. *Nature, Lond.* **201**, 638.
310. SCHMIDT-KOENIG, K. (1965). Current problems in bird orientation. *Adv. in the study of behaviour* **1**, 217–78.
311. SCHMIDT-NIELSEN, K., JØRGENSEN, C. B. and OSAKI, H. (1958), Extra-renal salt excretion in birds. *Am. J. Physiol.* **193**, 101–7.
312. SCHMIDT-NIELSEN, K. and SLADEN, W. J. L. (1958). Nasal salt secretion in the Humboldt penguin. *Nature, Lond.* **181**, 1217–8.
313. SCHOLANDER, P. F. (1940). Experimental investigations on the respiratory function in diving mammals and birds. *Hvalråd. Skr.* No. 22.
314. SCHÜZ, E. (1951). Uberblick über die Orientierungsversuche der Vogelwarte Rossitten (jetzt: Vogelwarte Radolfzell). *Proc. X int. orn. Congr.* 249–68.
315. SCHWARTZKOPF, J. (1963). Morphological and physiological properties of the auditory system of birds. *Proc. XIII int. orn. Congr.* **2**, 1059–68.
316. SERVENTY, D. L. (1956). Age at first breeding of the short-tailed shearwater *Puffinus tenuirostris. Ibis* **98**, 532–3.
317. SERVENTY, D. L. and MARSHALL, A. J. (1957). Breeding periodicity in Western Australian birds: with an account of the unseasonal nestings in 1953 and 1955. *Emu* **57**, 99–126.
318. SHANK, M. C. (1959). The natural terminaton of the refractory period in the slate-colored junco and in the white-throated sparrow. *Auk* **76**, 44–54.
319. SHANNON, J. A. (1938). The excretion of uric acid by the chicken. *J. cell. comp. Physiol.* **11**, 133–48.
320. SIBLEY, C. G. (1960). The electrophoretic patterns of avian egg-white proteins as taxonomic characters. *Ibis* **102**, 215–34.
321. SIMMONS, K. E. L. (1966). Anting and the problem of self-stimulation. *J. Zool.* **149**, 145–62.
322. SKUTCH, A. F. (1957). The incubation patterns of birds. *Ibis* **99**, 69–93.
323. SKUTCH, A. F. (1966). A breeding bird census and nesting success in Central America. *Ibis* **108**, 1–16.
324. SKUTCH, A. F. (1967). Adaptive limitation of the reproductive rates of birds. *Ibis* **109**, 579–99.

325. SMITH, C. HORTON- and AMOROSO, E. C. (1966). *Physiology of the Domestic Fowl*. Oliver & Boyd, Edinburgh.
326. SMITH, D. SUMMERS- (1958). Nest-site selection, pair formation and territory in the house-sparrow, *Passer domesticus*. *Ibis* **100**, 190–203.
327. SMITH, G. ELLIOT (1903). On the morphology of the brain in the Mammalia, with special reference to that of the lemurs, recent and extinct. *Trans. Linn. Soc. Lond.* 2nd series, **8**, 319–432.
328. SMITH, N. G. (1968). The advantage of being parasitized. *Nature, Lond.* **219**, 690–4.
329. SMITH, S. M. (1967). Seasonal changes in the survival of the black-capped chickadee. *Condor* **69**, 344–59.
330. SNOW, D. W. (1956). Territory in the blackbird *Turdus merula*. *Ibis* **98**, 438–47.
331. SOIKKELI, M. (1967). Breeding cycle and population dynamics in the dunlin (*Calidris alpina*). *Annls zool. fenn.* **4**, 158–98.
332. SOUTHERN, H. N. (1938). The spring migration of the willow-warbler over Europe. *Br. Birds* **32**, 202–6.
333. SOUTHERN, H. N. (1959). Mortality and population control. *Ibis* **101**, 429–36.
334. SOUTHERN, H. N. and LOWE, V. P. W. (1968). The pattern of distribution of prey and predation in tawny owl territories. *J. Anim. Ecol.* **37**, 75–97.
335. SOUTHERN H. N. and REEVE, E. C. R. (1941). Quantitative studies in the geographical variation of birds.—The common guillemot (*Uria aalge* Pont.). *Proc. zool. Soc. Lond.* A**111**, 264.
336. SPALDING, D. A. (1873). Instinct, with original observations on young animals. *Macmillan's Magazine* **27**, 282–93.
337. SPALDING, D. A. (1875). Instinct and acquisition. *Nature, Lond.* **12**, 507–8.
338. SPERBER, I. (1948). Investigations on the circulatory system of the avian kidney. *Zool. Bidr. Upps.* **27**, 429–48.
339. STAFFORD, J. (1969). The census of heronries, 1962–3. *Bird Study* **16**, 83–8.
340. STAGER, K. E. (1967). Avian olfaction. *Am. Zool.* **7**, 415–9.
341. STANTCHINSKI, V. V. (1927). Some climatic limits in the extension of birds in eastern Europe. *Ecology* **8**, 232–7.
342. STEEL, E. and HINDE, R. A. (1966). Effect of exogenous serum gonadotrophin (PMS) on aspects of reproductive development in female domesticated canaries. *J. Zool.* **149**, 12–30.
343. STENGER, J. (1958). Food habits and available food of ovenbirds in relation to territory size. *Auk* **75**, 335–46.
344. STENHOUSE, J. H. (1925). Mixed plumages in a brood of hybrid crows. *Scott. Nat.* **42**, 101–5.
345. STEVENSON, J. (1933). Experiments on the digestion of food by birds. *Wilson Bull.* **45**, 155–67.
346. STEWART, R. E. and ALDRICH, J. W. (1951). Removal and repopulation of breeding birds in a spruce-fir forest community. *Auk* **68**, 471–82.
347. STIEGLITZ, W. O. and THOMSON, R. L. (1967). Status and life-history of the Everglade kite in the United States. *Spec. scient. Rep. U.S. Fish Wildl. Serv.* **109**.
348. STONEHOUSE, B. (1962). The tropic birds (genus *Phaeton*) of Ascension Island. *Ibis* **103b**, 124–61.
349. STORER, R. W. (1945). Structural modifications in the hindlimb of the Alcidae. *Ibis* **87**, 433–56.

350. STORER, R. W. (1960). Evolution in the diving birds. *Proc. XII int. orn. Congr.* **2**, 694–707.
351. SUSHKIN, P. P. (1927). On the anatomy and classification of the weaver-birds. *Bull. Am. Mus. nat. Hist.* **57**, 1–32.
352. SVÄRDSON, G. (1949). Competition and habitat selection in birds. *Oikos* **1**, 157–74.
353. SWANK, W. G. (1955). Nesting and production of the mourning dove in Texas. *Ecology* **36**, 495–505.
354. SWINTON, W. E. (1958). *Fossil Birds.* British Museum (Natural History), London.
355. SWYNNERTON, C. F. M. (1915). Mixed bird parties. *Ibis* **57**, 346–54.
356. SYKES, A. H. (1966). Renal function in fowl. In HORTON-SMITH and AMOROSO (ref. 325).
357. TANSLEY, K. (1965). *Vision in Vertebrates.* Chapman and Hall, London.
358. TAYLOR, T. G. (1966). The endocrine control of calcium metabolism in fowl. In HORTON-SMITH and AMOROSO (ref. 325).
359. TEAGER, C. W. (1967). Birds sunbathing. *Br. Birds* **60**, 361–3.
360. THOMPSON, C. NETHERSOLE- and THOMPSON, D. NETHERSOLE- (1943–44). Nest-site selection by birds. *Br. Birds* **37**, 70–4; 88–94; 108–113.
361. THORPE, W. H. (1951). The learning abilities of birds. *Ibis* **93**, 1–52; 252–96.
362. THORPE, W. H. (1961). *Bird Song.* Cambridge University Press, Cambridge.
363. THORPE, W. H. (1964). The isolate song of two species of *Emberiza*. *Ibis* **106**, 115–8.
364. THORPE, W. H. and GRIFFIN, D. R. (1962). Ultrasonic frequencies in bird song. *Ibis* **104**, 220–7.
365. THORPE, W. H. and PILCHER, P. M. (1958). The nature and characteristics of sub-song. *Br. Birds* **51**, 509–14.
366. TICEHURST, C. B. (1935). On the food of the barn owl and its bearing on barn owl population. *Ibis* **77**, 329–35.
366a. TICEHURST, C. B. (1939). On the food and feeding-habits of the long-eared owl (*Asio otus otus*). *Ibis* **81**, 512–20.
367. TINBERGEN, N. (1951). *The Study of Instinct.* Clarendon Press, Oxford.
368. TOMPA, F. S. (1967). Reproductive success in relation to breeding density in pied flycatchers, *Ficedula hypoleuca* (Pallas). *Acta zool. fenn.* **118**, 3–28.
369. TUCKER, B. W. (1944). The ejection of pellets by passerine and other birds. *Br. Birds* **38**, 50–2.
370. TUCKER, B. W. (1949). Species and subspecies: a review for general ornithologists. *Br. Birds* **42**, 129–34; 161–74; 193–205.
371. TUCKER, D. (1965). Electrophysiological evidence for olfactory function in birds. *Nature, Lond.* **207**, 34–6.
372. TUCKER, V. A. (1966). Oxygen consumption of a flying bird. *Science, N.Y.* **154**, 150–1.
373. TUNMORE, B. G. (1960). A contribution to the theory of bird navigation. *Proc. XII int. orn. Congr.* **2**, 718–23.
374. TURČEK, F. J. (1949). The bird population in some deciduous forests during a gypsy-moth outbreak. *Zpr. st. vyzk. ust. les.* (*Bull. Inst. For. Res. Czechoslovakia*), 108–31.
375. TURČEK, F. J. (1958). The proportions of plumage, organic matter and water content in the bodies of some birds. *Proc. XII int. orn. Congr.* **2**, 724–9.

376. UDVARDY, M. D. F. (1953). Contributions to the knowledge of the body temperature of birds. *Zool. Bidr. Upps.* **30**, 25–32.
377. UDVARDY, M. D. F. (1956). Observations on the habitat and territory of the chaffinch *Fringilla c. coelebs*, in Swedish Lapland. *Ark. Zool.* **9**, 499–505.
378. UDVARDY, M. D. F. (1963). Data on the body temperature of tropical sea and water birds. *Auk* **80**, 191–4.
379. VERBECK, N. A. M. (1964). A time and energy study of the Brewer blackbird. *Condor* **66**, 70–4.
380. WALLGREN, H. (1954). Energy metabolism of two species of *Emberiza* as correlated with distribution and migration. *Acta zool. fenn.* **84**, 1–110.
381. WARD, R. (1966). Regional variation in the song of the Carolina chickadee. *Living Bird* **5**, 127–50.
382. WATERTON, C. (1832). On the faculty of scent in the vulture. *Loudon's Mag. nat. Hist.* **5**, 233–41.
383. WATSON, A. (1964). Aggression and population regulation in red grouse. *Nature, Lond.* **202**, 506–7.
384. WATSON, A. (1965). Research on Scottish ptarmigan. *Scott. Birds* **3**, 331–49.
385. WATSON, A. and JENKINS, D. (1968). Experiments on population control by territorial behaviour in red grouse. *J. Anim. Ecol.* **37**, 595–614.
386. WEISE, C. M. (1956). Nightly unrest in caged migratory sparrows under outdoor conditions. *Ecology* **37**, 274–87.
387. WEISE, C. M. (1962). Migratory and gonadal responses of birds on long-continued short day-lengths. *Auk* **79**, 161–72.
388. WEISE, C. M. (1963). Annual physiological cycles in captive birds of differing migratory habits. *Proc. XIII int. orn. Congr.* **2**, 983–93.
389. WEISE, C. M. (1967). Castration and spring migration in the white-throated sparrow. *Condor* **69**, 49–68.
390. WEIS-FOGH, T. (1967). Metabolism and weight economy in migrating animals, particularly birds and insects. In *Insects and Physiology*, ed. BEAMENT, J. W. L. and TREHERNE, J. E., Oliver and Boyd, Edinburgh.
391. WENZEL, B. M. (1968). Olfactory prowess of the kiwi. *Nature, Lond.* **220**, 1133–4.
392. WERTH, IRENE. (1960). The problem of flocking in birds. *Proc. XII int. orn. Congr.* **2**, 744–8.
393. WEST, G. S. (1965). Shivering and heat production in wild birds. *Physiol. Zoöl.* **38**, 111–20.
394. WETMORE, A. (1960). A classification for the birds of the world. *Smithson. misc. Collns* **139** (11), 1–37.
395. WILKINSON, D. H. (1952). Randomness in bird navigation. *J. exp. Biol.* **29**, 532–60.
396. WILLIAMS, C. B. (1951). Intergeneric competition as illustrated by Moreau's records of east African bird communities. *J. Anim. Ecol.* **20**, 246–53.
397. WINTERBOTTOM, J. M. (1943). On woodland bird parties in northern Rhodesia. *Ibis* **85**, 437–42.
398. WITSCHI, E. (1936). Effect of gonadotropic and oestrogenic hormones on regenerating feathers of weaver-finches (*Pyromelana franciscana*). *Proc. Soc. exp. Biol. Med.* **35**, 484–9.
399. WITSCHI, E. (1961). Sex and secondary sexual characters. In MARSHALL (ref. 4).

400. WOLFSON, A. (1958). Regulation of refractory period in the photoperiodic responses of the white-throated sparrow. *J. exp. Zool.* **139**, 349–80.

401. WOLFSON, A. (1959). The role of light and darkness in the regulation of spring migration and the reproductive cycle in birds. In *Photoperiodism and Related Phenomena in Plants and Animals*, ed. WITHROW, R. B., American Association in the Advancement of Science. Bailey Bros., London.

402. WOLFSON, A. (1959). Role of light in the photoperiodic responses of migratory birds. *Science, N.Y.* **129**, 1425–6.

403. WOLFSON, A. (1960). The ejaculate and the nature of coition in some passerine birds. *Ibis* **102**, 124–5.

404. WOLFSON, A., GOLDBERG, S. M., TOMM, K. E. and WESTERHOFF, T. R. (1960). Gonadal response to exogenous gonadotropin during the regressive phase of the annual cycle in passerine birds. *Anat. Rec.* **137**, 401–2.

405. YAPP, W. B. (1951). The population of rooks (*Corvus frugilegus*) in west Gloucestershire. II. *J. Anim. Ecol.* **20**, 169–72.

406. YAPP, W. B. (1962). *Birds and Woods*. Oxford University Press, London.

407. YAPP, W. B. (1963). Color variation and status of *Parus ater britannicus* and *P. a. hibernicus*. *Proc. XII int. orn. Congr.* **1**, 199–201.

408. YAPP, W. B. (1969). The bird population of an oakwood (Wyre Forest) over eighteen years. *Proc. Bgham nat. Hist. Soc.* **21**, 199–216.

409. YARRELL, W. (1829). On the structure of the beak and its muscles in the crossbill, (*Loxia curvirostra*). *Zool. J.* **4**, 459–65.

410. ZUCKERMAN, S. and SUDERMAN, A. E. (1935). Serum relationships within the family Cercopithecidae. *J. exp. Biol.* **12**, 222–8.

Index

Bold type indicates an illustration. Where there is more than one reference *italic* indicates importance.

Accipiter nisus, 157, 203
accommodation, 87
A-cells, 70
acetylcholine, 30
Acrocephalus palustris, 135, 208
  schoenobanus, 134, 208
  scirpaceus, 134, 208
acuity, 89
adaptation, 13
adaptive radiation, 44–58
adenohypophysis, 96, 99
adrenals, 153
Aegithalos caudatus, 118, 207
Aegithognathae, 41
Aepyornis, 46, 202
Aepyornithiformes, 202
Agelaius phoeniceus, 73, 207
  tricolor, 122, 207
air-sacs, 21, **34**, 35, **36**, 76
Aix sponsa, 192, 203
Alauda arvensis, 127, 206
Alaudidae, 206
albatross, 22, 28, 86, 186
  Laysan, 190, 202
  Royal, 94, 202
albumen, 6
albumin, 6
Alca torda, 54, 204
Alcae, **43**, 51, 118, 204
Alcedidae, 118
Alcedo atthis, 140, 205
alimentary canal, **66**–67
alula, 28, **29**
ambiens muscle, 23, **24**, 41
aminoacids, 65
Ampullaria paludris, 189
Anas platyrhynchos, 74, 203
ancestry, 1
androgens, 97, 99, 107
Anhimae, 203
ankle, 18, **19**
Anomolognathae, 41
Anseres, 86, 118, 203
Anseriformes, 51, 203
Anthus pratensis, 198, 207
  trivialis, 138, 207
apnoea, 74
Apodemus flavicollis, 110
  sylvaticus, 110
Apodi, 205
Apodidae, 118
Apodiformes, 56, 205
aposematic display, 117
Aptenodytes forsteri, 117, 202
  patagonica, 118, 202
apteria, 32, 123
Apterygiformes, 202
Apteryx, 46
  australis, 202
Apus apus, 119, 184, 205
  melba, 184, 205
Aquila chrysaetos, 119, 203

Arachis hypogaea, 110
Archaeopteryx lithographica, **2**, *3–4*, **3**, 13, 20, 32, 41, 201
Archaeornithes, 41, 201
Archilochus alexandri, 73, 205
colubris, 146, 205
Archosauria, 1
Ardea cinerea, 173, 203
area, 88, 89
arterial arches, **11**
Asio flammeus, 68, 205
otus, **56**, 67, 205
aspect ratio, 28
Atlantisia rogersi, **45**, 203
auk, great, 45, 55, 204
auks, 42, 52, 54, 118, 166, 204
autonomic system, 83
Aves, 41, 201
avocet, 51, 204

Baer, 42
balance, 38
barbules, 24
basilar membrane, 91
bastard wing, 23, 28, **29**, 47
bat, 13, 15
bathing, 171
B-cells, 70
beak, 106–07
bee-eater, red-throated, 164, 205
bee-eaters, 118, 152, 164
behaviour, 42, 80, 107–08
bile-salts, 70
binocular vision, 89
bipedality, 17–19
birds of prey, 8, 55, 68, 180
birth rate, 183
blackbird, 69, 111, 119, **130**, 132, **138**, 139, 140, 141, 142, 170, 184, 191, 196, 198, 209
red-winged, 73, 207
blackcap, 127, 134, 146, **148**, 151, 196, 208
blackcock, 116, 203
blastodisk, 6
blood system, 11–12
bluethroat, 151, 209
bobolink, 147, 207
Bookham Common, 178
bower-bird, Satin, 208
bower-birds, 116
Bowman's capsule, 9
bradycardia, 75
brain, 38, **39**, **83**, **84**
brambling, 127, **131**, 198, 206
breathing, 35
breeding seasons, 94–104
broadbills, 205
bronchus, 35
brood parasites, *124–25*, 186
patches, 97, 103, 123, 124
brooding, 97, 103, 123
Bubo virginianus, 68, 205
Buceros, 69
bicornis, 205
budgerigar, 30, 94, 204
bullfinch, 32, 67, 164, 206
bunting, corn, **129**, 137, 141, 206
reed, **138**, 198, 206
buntings, 67, 127, 142
bush-tit, common, 207
bush-tits, 123
bustard, 51, 203
Buteo buteo, 180, 203
buzzard, 22, 180, **182**, 203

Cactospiza pallida, 166, 206
caecum, 46, 66, 67, 69
calcium, 107
Calidris alpina, 192, 204
call-notes, **138**, 139
Calypte costae, 73, 205
Camarhynchus pallidus, 166, 206
Campylorhynchus brunneicapillus, 20[illegible]
canary, 97, 122, 123, 132
Capella gallinago, 204
capon, 106
Caprimulgiformes, **43**, 55, 118, 205
carbon, biochemistry of, 70–71
cardinal, 76, 77, **78**, **79**, 206
Carduelis carduelis, 110, 206
flammea, 94, 206
carina, 16
Carinatae, 40, 41, 46, 202

carnivores, 67
carotenoid, 33
carotid bodies, 73
carpometacarpus, 4, 14, 19
cassowary, 46, 47, **48**, 49, 202
castration, 106, 107, 153
Casuariiformes, 202
Casuarius, 46, 48
  casuarius, 202
Cathartae, 203
Cathartes aura, 86, 203
censuses, 172
Centrocercus urophasianus, 116, 203
Centropelma micropterum, 45, 202
Cepaea, 165
cerebellum, 83, 84
cerebral hemispheres, 38, 83
Certhia familiaris, 167, 206
Certhiidae, 206
chaffinch, **28**, 59, 63, 64, 67, 104, 107, 111, 115, 120, 127, 132, 135, **136**, **137**, **138**, 139, 141, 145, **146**, 152, 153, 186, 191, 197, 198, 206
  blue, 63–64, 206
chalaza, 6
Charadrii, **43**, 86, 204
Charadriiformes, 42, **43**, 51, 204
Charadrius hiaticula, 124, 204
Chauna torquata, **21**, 203
chemical senses, 85–87
chickadee, black-capped, 73, 188, 207
  boreal, 197
  brown-capped, 197, 207
  Carolina, 136, 207
  chestnut-backed, 197, 207
  grey-headed, 207
chiffchaff, 127, **134**, 135, 151, 208
Chiroxiphia, 116
  pareola, 205
chitinase, 69
Chloris chloris, 206
Chordeiles minor, 73, 203
Ciconia ciconia, 157, 203
Ciconiiformes, 55, 203
Cinclidae, 206
Cinclus, 55
  cinclus, 140, 205
classification, 40–64
clavicle, 15, 22
cleidoic egg, 7
climate, 193
cline, 59–61
clutch size, 183–86
cnemial crest, 53
Coccothraustes coccothraustes, 117, 206
cochlea, **90**
coelomoduct, 9
Coliiformes, 205
Colius, 81
  colius, 205
collar bone, 15
Collocalia, 38, 118
  vestita, 205
colonies, 166–67
colour vision, 89, 115
colours of eggs, 7
  of feathers, 32–33
Columba, **14**, **18**, **24**
  oenas, 204
  palumbus, 67, 204
Columbiformes, 44, 204
columella auris, 90
Colymbus immer, 53
comb, 106–07
combassou, 124, 139
  Senegal, 208
Compsothlypidae, 208
conditioned reflex, 83, 89, *112–13*
condor, 94
  Andean, 203
  Californian, 203
cones, 88
coniferous plantations, 191
coot, 45, 51, **52**, 204
copulation, 93
Coraciiformes, 205
coracoid, 15
cormorant, common, 202
  flightless, 45, 202
cormorants, 21, 51, **52**, 72, 74, 76, 123
corncrake, 175, 204
cortex of cerebral hemisphere, 38
  of gonad, 92
Corti's organ, 90
Corvidae, 189, 206

Corvus brachyrhynchos, 157, 206
corax, 91
corone cornix, 62, 152, 206
corone corone, 62, 152, 206
frugilegus, 68, 206
monedula, 91, 206
Coturnix coturnix, 94, 203
courtship, 114–16
coverts, 24
cowbird, 186
bay-winged, 124, 207
brown-headed, 124, 207
giant, 190, 207
crane, 51
whooping, 172, 203
Crex crex, 175, 204
critical temperatures, 77
crocodile, 11, 20
crop, 67
crossbill, 61, 72, 94, 164, 206
crow, American, 157, 206
carrion, 62, **63**, 152, 153, 206
hooded, 62, **63**, 152, 153, 157, 206
crows, 39, 58, 106, 166
cryptic coloration, 33, 104
Crypturellus soui, 202
cubitals, 23
cuckoo, 28, 124, 125, 127, 156, 170, 191, 204
Cuculidae, 124
Cuculiformes, 204
Cuculus canorus, 124, 204
curlew, 169, 191, 204
Cyanosylvia suecica, 151, 209
Cynuris coccinigaster, 207

Darwin, 45, 185
Dean Forest, 145
death rate, 187
decreases in numbers, 175–81
Delichon urbica, **15**, **19**, 118, 207
Dendrocopus major, 124, 205
minor, 167, 205
Dendroica kirtlandi, 172, 208
density-dependent factors, 187
Desmognathae, 41
digestion, 67–70
digitals, 23
dimorphism, 104, 106, 126
Dinornis, 46, 202
Dinornithiformes, 202
dinosaurs, 1, 19, 47
dioch, red-billed, 102, 103, 208
Diomedea epomophora, 94, 202
immutabilis, 190, 202
dipper, 55, 140
Dipus, 19
disease, 187, 190
dispersion, 172–200
colours, 33
distribution, 172–200
divers, 51, 202
diving, 74–75
diving birds, 49–55
dodo, 17, 190, 204
Dolichonyx orizivorus, 147, 207
dove, mourning, 80, 94, 204
ringed turtle, 107, 204
stock, 185, 204
doves, 127
down, 32, 104
Dreissena polymorpha, 189
Dromaeognathae, 41
dromaeognathous palate, 46
Dromaius, 46
novaehollandiae, 202
drumming, 126
duck, steamer, 45, 203
wood, 192, 203
ducks, **17**, 21, 30, 35, 51, 62, 71, 73, 74, 75, **85**, 86, 92, 93, **99**, 106, 118, 189
duet, 132
dunlin, 192, 204
dusting, 171
eagle, 22
golden, 119, 188, 203
ear, 55, **56**, 127
echolocation, 38
eclipse plumage, 106
ectoparasites, 171
Ectopistes migratorius, 190, 204
eggs, 5–8
electrophoresis, 42, **43**, 47
elephant bird, 46, 202

emarginated primaries, 28
Emberiza calandra, **129**, 206
  citrinella, 132, 206
  hortulana, 154, 206
  schoeniclus, 198, 206
Emberizidae, 127, 201
Empidonax, 123
  traillii, 205
emu, 41, 46, 47, 86, 202
endocrine control, 92
energy, production of, 72–74
epidermal structures, 106–07
epigamic display, 117
Erithacus rubecula, 32, 88, 209
Ernarmonia conicolana, 163
Eryops, 20
Estrilda troglodytes, 73, 208
Estrildidae, 124, 201
ethology, 111
Eugenes fulgens, 73, 205
Eurylaimi, 205
evaporation, 76
exoskeleton, 8
eye, 87, 99

Falco aesalon, 189, 203
  peregrinus, 119, 203
  subbuteo, **56**, 203
  tinnunculus, 170, 203
falcon, peregrine, 119, 180, 203
Falcones, 203
Falconiformes, 55, 93, 203
falcons, 30, 55
fat, as fuel, 30, 74
  deposition of, 153, 154
fatigue, 112
feather tracts, **31**
feathers, 8, *23–24*, **25**, 75, *104–06*
  maintenance of, 171
feet, **50**, **52**, **56**
fertilization, 93
Ficedula albicollis, 207
  hypoleuca, 207
fieldfare, 193, 209
finch, zebra, 94, 208
finches, 67, 72, 76, 93, 100, 102, 132, 140, 143, 169

firefinch, 139, 186
  Senegal, 208
flapping flight, 26
flight, 13–39
flightless birds, 5, *45–49*, 190
flocking, 166–68, **168**
fluttering, gular, 76, 80
flycatcher, alder, 205
  American, 123
  pied, 117, 120, **121**, **133**, 142, 143, **144**, 145, 190, 192, 196, 198–99, 207
  spotted, 167, 196, 207
  white-collared, 157, 207
follicle-stimulating hormone (FSH), 96, 97
food, 65–66
  seeking, 163–65
  storage of, 165
foreplay, 93, 115, 116, 117
fovea, 88, 89
fowl, 8, **34**, 35, **36**, 69, 73, 76, 80, 83, 85, 93, 98, 101, 106, 107, 112, 132
Fratercula arctica, 54, 204
Fregata, 15, 202
frigate-birds, 15, 76, 202
Fringilla coelebs, 28, 59, 63, 64, 206
  montifringilla, 127, 206
  teydea, 63, 64, 206
  tintillon, 64
Fringillidae, 206
Fulica atra, 51, 204
fulmar, 28, 94, 166, 181, 183, 199, 202
Fulmar glacialis, 28, 202
furcula, 15, 16

Gadow, 42, 201
Galapagos woodpecker-finch, 166, 206
Galliformes, 203
Gallinula, 51
  chloropus, 204
game-birds, 8, 22, 38, 66, 67, 75, 76, 93, 98, 106, 116, 171, 183, 187
gamosematic display, 115

gannet, 51, 76, 104, 122, 123, 160, 166, 172, 186, 202
garden birds, 198
Garrulus glandarius, 165, 206
Gause, 196
Gaviiformes, 51, 202
geese, 8, 51, 62, 69, 70, 85, 86, 93, 118
genotype, 111
germinal disk, 6
gizzard, 67, 69
gliding, 27–28
glomerulus, 9, 71
glossy starling, 33
glucagon, 70
goldcrest, 119, 208
goldfinch, 110, 113, 115, 206
gonadotropins, 96, 99, 153
gonads, *92–94*, 153
Grandry corpuscles, **85**
grebe, flightless, 45, 202
  great crested, **52**, 74, 175, 176, **177**, 202
grebes, 45, 51, 53, 124, 202
greenfinch, **16**, 206
greyhen, 116, 203
Grinnell, 196
grouse, red, 142, 145, 189, 192, 203
  sage, 116, 203
Gruidae, 51, 203
Gruiformes, 51, 52, 203
Grus americana, 172, 203
guanine, 71
guillemot, **52**, **60**, 61, 166, 204
gular fluttering, 76, 80
gull, black-headed, 8, 105, 106, 107, 124, 167, 204
  glaucous, **50**, 204
  great black-backed, 73, 204
  herring, 106, 167, 196, 204
  lesser black-backed, 167, 196, 204
  western, 80, 204
gulls, 28, 42, **43**, 51, 68, 94, 104, 144, 166
Gymnogyps, 94
  californicus, 203

habitat selection, 193–95
habituation, 112
Haematopus ostralagus, 123, 204
haemoglobin, 73, 75
Handley Page slots, 28
hawfinch, 117, 141, 164, 206
hawks, 38, 55, 72, 88, 89, 203
hearing, 37–38, *89–91*
heart, 11
Helix, 165
Henle, loop of, 9
Herbst corpuscle, **85**
heron, common, 203
herons, 19, 55, 76, 173, 200
Hesperornis, 4, 41, 201
Hesperornithiformes, 201
hibernation, 81
Hirundinidae, 118, 207
Hirundo, 152
  rustica, 147, 207
hoatzin, **8**, 67
hobby, **56**, 203
hole-nesting, 140, 142, 184
homing, 156, *160–62*
homoiothermy, 32, 75
Homolognathae, 41
honey-guide, black-throated, 205
honey-guides, 69, 124
hooks, 24
hormones, 96–108
hornbill, 69
  great, 205
Howard, Eliot, 129
humerals, 23
humerus, 22
humming-bird, bee, 205
  black-chinned, 205
  blue-throated, 205
  Costa's, 205
  Rivoli's, 205
  ruby-throated, 146, 205
humming-birds, 16, 22, 33, 56, 67, 72, 73, 81, 89
Hydrobia ulvae, 164
hyperdactyly, 55
hyperglycaemia, 70
hyperphalangy, 55
Hypochaera, 139
  chalybeate, 208

hypothalamus, 99

Ichthyornis, 4, 41, 202
Ichthyornithiformes, 202
ichthyosaur, 49
Icteridae, 118, 190, 207
Iguanodon, 19
immune reactions, 42
Impennes, 41, 52, 202
increases in population, 181–83
Indicator indicator, 205
Indicatoridae, 69, 124
instinct, *109–14*, 163
insulin, 70
intelligence, 109–14
interclavicle, 15
interference colours, 33
intersexes, 93
interstitial-cell stimulating hormone (ICSH), 96
intestine, 67, 69
islets of pancreas, 70
isolation, 62

jackdaw, 91, 119, 206
jay, 165, 198, 206
  Canadian, 78, 206
Junco hyemalis, 92, 206
junco, slate-colored, 92, 98, 102, 103, 206
juvenile plumage, 104
Jynx, 19
  torquilla, 179, 205

kakapo, 17, 205
keel, 16, 22, 46
keratin, 6, 8, 33
kestrel, 170, 203
kidney, 9–10, 71–72
kinetism, 46
kingfisher, 33, 49, 89, 118, 122, 199
  common, 140, 205
kite, Everglade, 189, 203
kittiwake, 167, 204
kiwi, 46, 47, 49, 86
Kramer, 157

Lagonostricta, 139
  senegala, 208
Lagopus lagopus, 142, 203
  mutus, 33, 203
Lampornis clemencii, 73, **82**, 205
Laniidae, 207
Lanius collurio, 117, 207
lapwing, 28, 204
Lari, **43**, 204
Laridae, 51
larks, 140, 171
Larus argentatus, 106, 204
  fuscus, 167, 204
  hyperboreus, 204
  marinus, 73, 204
  occidentalis, 80, 204
  ridibundus, 105, 204
Laspeyresia conicolana, 163
leaf-warblers, 118
learning, *111–13*, 163, 171
  to sing, 136–39
lecithin, 5
lecithovitellin, 5
lek, 116, 144
Lemur, **84**
Leydig cells, 107
lice, 42, 47
light and breeding, 97–104
  and migration, 153
Liparis dispar, 199
Locustella naevia, **128**, 208
loons, 51, 53, 75, 202
loss of heat, 76
loudness, 127
Loxia curvirostra, 62, 206
lungs, **34**, 35, **36**, 76
Luscinia megarhynchos, 127, 209
luteinizing hormone (LH), 96, 97
luteotropic hormone (LTH), 96
lyrebirds, 58, 205
Lyrurus tetrix, 116, 203

macula, 88
magnetism, 161

magpie, 198, 206
mallard, 157, 162
Mallophaga, 42
manakin, 116, 205
Maniola jurtina, 59
marrow, 107
martin, house, **15**, **19**, 118, 119, 124, 154, 198, 207
sand, 118, 192, 198, 207
Mayr, 59
meadow brown, 59
mechanics of flight, 25–29
median eminence, 99
medulla of gonad, 92, 93
Megapodidae, 124
melanin, 33
Meleagris gallopavo, **20**, 93, 203
Mellisuga helenae, 56, 205
Melopsittacus undulatus, 30, 204
Melospiza lincolni, 101, 206
melodia, 62, 206
Menurae, 205
Menuridae, 58
merlin, 189, 203
Meropidae, 118
Merops, 152
bullocki, 164, 205
Merrem, 46
metabolic rate, 72, 105
metabolism, 70–75
metacarpal quills, 23
mice, 110
Microtus agrestis, 181
migration, 145–62
Mimidae, 207
Mimus polyglottus, 135, 207
moa, 46, 47, 202
mobbing, 169–70
mocking bird, 135, 207
Moffat, 139
Molothrus ater, 124, 207
badius, 124, 207
Montifringilla montifringilla, 206
moorhen, 45, 51, 124, 204
mortality, 187–90
Motacilla aguimp, 152, 207
flava, 152, 207
Motacillidae, 207
moult, 33, 104
mouse-bird, 81
white-backed, 205
Müllerian duct, 93
mimicry, 139
Mus musculus, 110
Muscicapa albicollis, 157, 207
hypoleuca, 117, 207
striate, 207
Muscicapidae, 201, 207
muscles, 22–23, 51
muscular contraction, 30
myoglobin, 75
Mytilus edulis, 165
myxomatosis, 181

Nannopterum harrisi, 45, 202
nasal gland, 72
navigation, 156
Nectariniidae, 105, 207
neopallium, 38
Neornithes, 41, 201
nephron, 9
nerve-muscle junction, 30
nervous system, 81–84
nest building, 120–22
sanitation, 122
site, *118–20*, 190, 193, 199
nest-boxes, 142, 192
nesting, 8, *117–22*
neurohumors, 99
Newton, 44, 58
niche, 196
nidicolous birds, 80
nidifugous birds, 80
nighthawk, common, 73, 79, 80, 118, 198, 205
nightingale, 127, **130**, 151, 193, **194**, **195**, 209
nightjars, 38, 42, **43**, 55, 76, 88, 118
nitrogen, biochemistry of, 71
noise, 128
non-shivering thermogenesis, 77
Nucifraga caryocatactes, 165, 206
numbers, 172–200
Numenius arquata, 169, 204
numerical taxonomy, 41

nutcracker, 165, 206
nuthatch, 61, 165–66, 167, 208
  brown-headed, 166, 208
nutrition, 65–70
nuts, ground, monkey or pea, 110

Odontognathae, 41, 201
Oenanthe oenanthe, 33, 209
oestradiol, 123
oestrogens, 96, 97, 99, 103, 107, 123, 142
oilbird, 38, 86, 205
olfactory lebes, 83, 86
Opisthocomus hoazin, 8, 203
optic lobes, 83
ornithine cycle, 71
Ornithischia, 1
Ornithosuchus, **2**
ortolan, 154, 206
Oscines, 57, 58, 126, 205
osprey, 180, 203
ostrich, 8, 41, 46, 47, **48**, 49, 89, 91, 106, 118, 184, 202
Otididae, 51, 203
ovary, 92, 93, 96
ovenbird, 199, 208
oviduct, 93, 101
ovomucoid, 6
Owen, 4
owl, barn, **12**, **20**, 68, 197, 205
  great horned, 68, 69, 205
  long-eared, **56**, 67, 67, 205
  short-eared, 68, 184, 205
  tawny, 132, 141, 169, 188, 205
owls, 15, 38, 42, **43**, 55, 68, 70, 76, 88, 90, 91, 184, 188
oxygen debt, 74, 75
oyster-catcher, 123, 165, 204

pain, 84
pair formation, 114–17
palaeognathous palate, 46
palaeopallium, 39
palate, 46
pancreas, 70
Pandion haliaetus, 180, 203
panting, 76
Panurus biarmicus, 191, 208
parabronchus, 35
Paradoxornithidae, 208
parasites, 42, 187, 190
parathyroid, 107
Paridae, 150
parrot, grey, 91, 204
parrots, 15, 30, 38, 57, 58, 67, 69, **83**
partridge, 32, 203
Parulidae, 208
Parus ater, 62, 207
  atricapillus, 73, 207
  caeruleus, 33, 207
  carolinensis, 136, 207
  cinctus, 165, 207
  cristatus, 165, 207
  hudsonicus, 197, 207
  inornatus, 141, 207
  major, 104, 207
  montanus, 118, 207
  palustrus, 118, 207
  rufescens, 197, 207
Passer domesticus, 42, 208
  montanus, 199, 208
Passerculus sandvicensis, 72, 206
Passerella iliaca, 103, 206
Passeres, 205
Passeriformes, passerines, 44, 57, 93, 110, 157, 171, 183, 205
patagium, patagial membrane, 22, 23, 24
patella, 53
Pavlov, 112
Pavo cristatus, 115, 203
peacock, peafowl, 33, 115, 203
peck-order, 169, 187
pecten, 88
pectoral muscle, pectoralis major and minor, 22, **23**
pelican, 42, 51, 76, 169
  brown, 202
Pelicaniformes, 51, 202
Pelicanus, 169
  occidentalis, 202
pellets, 68
pelvic girdle, 17

penguin, 41, 42, **43**, 51, **52**, **54**, 55, 72, 74, 94, 202
Adelie, 160, 202
emperor, 117, 202
king, 117, 202
penis, 92, 93
perching birds, 57
Perdix perdix, 33, 203
peregrine, 15
Perisoreus canadensis, 78, 206
pesticides, 180
petrel, 42, 51, 94, 118, 166
Petrochelidon pyrrhonota, 118, 207
Phaeton aethereus, 96, 202
lepturus, 96, 202
Phalaenopterus nuttalli, 76, 205
phalarope, 51
grey, red, red-necked, **50**, 204
Phalaropidae, 51
Phalaropus fulicarius, 204
Phalocrocorax carbo, 74, 202
Phasianus colchicus, 203
pheasant, 22, 32, 86, 203
Reeves', 107, 203
phenotype, 111
Philomachus pugnax, 104, 204
Philornis, 190
Phoenicurus phoenicurus, 135, 209
phosphate, 107
photoperiods, 153
Phylloscopus, 118
collybita, 127, 208
sibilatrix, 208
trochilus, 208
Pica pica, 198, 206
Picidae, 118
Piciformes, 55, 205
Picoides tridactylus, 47, 205
Picus viridis, 198, 205
pigeon, **14**, **18**, 22, 30, 33, 35, 38, **39**, 66, 67, 69, 76, 85, 86, **90**, 93, 106, 116, 117, 126, 156, 160, 161
passenger, 190, 204
pigeon's milk, 67
Pinguinis impennis, 45, 204
pipit, meadow, 191, 198, 207
tree, 138, 140, 152, 193, 198, 207
pitch, 127, 128
pituitary, 96, **100**, 153
Ploceidae, 44, 105, 106, 118, 124, 201, 208
plover, American golden, 147, **150**, 204
ringed, 124, 204
plovers, 22
plumage, 104–06
Pluvialis dominica, 147, 204
pneumatization, 21–22, 55
Podiceups cristatus, **53**, 74, 202
Podicipedidae, 124
Podicipediformes, 51
polygamy, 116
Pomacea paludris, 189
Pooecetes gramineus, 206
poorwill, 76, 81, 205
porphyrin, 6, 33
portamento, 128, 130
predation, 167, 187, 188, *189*
preening, 171
premigratory restlessness, 153, 157
prevomer, 46
primaries, 23, 24
Procellariiformes, **43**, 52, 86, 118, 202
procoracoid, 15, 22
production of heat, 77–80
progesterone, 97
progestin, 97
prolactin, 67, 96, 97, 103, 107, 123
promiscuity, 117
proventriculus, 67
Prunella vulgaris, 191, 208
Prunellidae, 208
Psaltriperus, 123
minimus, 207
Pseudosuchia, 1, 19, 47
Psittaciformes, 57, 58, 204
Psittacus erithacus, 91, 204
ptarmigan, 33, 192, 203
Pteroclididae, 44
Pterocnemia, 46
pennata, 202
pterodactyl, 13, **14**
Pteropus, 15
pterylae, **31**, 32
pterylosis, 46
Ptilonorhynchidae, 116, 208

Ptilonorhynchus violaceus, 208
puffin, 54, 204
Puffinus puffinus, 160, 202
  tenuirostris, 102, 202
Pygoscelis adeliae, 160, 202
pygostyle, 19, **20**
Pyrrhula pyrrhula, 32, 206
Pyrrhuloxia cardinalis, 76, 206

quail, 94, 203
Quelea quelea, 102, 208

rail, flightless, 45, 203
rails, 16, 45, 51, 203
rainfall and breeding, 102, 103
Rallidae, 45, 51, 203
Rallus madagascariensis, **45**, 204
Raphus cucullatus, 17, 204
raptors, 68, 164
Ratitae, ratites, 15, 16, 17, 40, *46–49*, 58, 67, 93, 202
raven, **57**, 91, 94
razorbill, **43**, 54, 204
Recurvirostra, 51
  avosetta, 204
redpoll, 94, 206
redshank, **50**, 204
redstart, 135, 184, 209
  American, 192, 208
redwing, 209
  tricolored, 122, 207
reeve, 116, 204
refractory period, 100, 103
Regulidae, 201, 208
Regulus regulus, 119, 208
remiges, 23, 55
renal portal system, 9, **10**
reproduction, 92–108
reproductive organs, 92–94
respiratory system, 35–37
retina, 88
rhea, 41, 46, 47
  common, 202
  Darwin's, 202
Rhea, 46
  americana, 202

Rheiformes, 202
rhinencephalon, 99
ribs, 19–21, **21**
Richmondena cardinalis, 76, **78**, 206
Riparia riparia, 118, 206
Rissa tridactyla, 167, 204
robin, 32, 88, 101, 111, 115, 126, **131**, 132, 141, 144, 158, 184, 191, 199, 209
Rodentia, 110
roding, 140
rods, 88
rook, 68, 94, 101, 166, 174, 206
roosting, 168–69
Rostramus sociabilis, 189, 203
Rowan, 92
ruff, 104, 106, 116, 204

saliva, 67
salivary glands, 67
sandgrouse, 44
  Pallas's, 204
Saurornithes, 41, 201
Saxicola rubetra, 198, 209
  torquata, 180, 209
scala media, 90
  vestibuli, 90
scales, 8
Scaphidura orizivora, 190, 207
scapula, 15
Schizognathae, 41
Sciurus carolinensis, 163
Scolopax rusticola, 140, 204
screamer, crested, 203
screamers, 21, 203
sea birds, 89, 94, 96, 147, 166, 184, 186
secondaries, 23, 24
secretin, 70
Seiurus aurocapillus, 199, 208
self-stimulation, 171
sense organs, 84–91
senses, 37–38, *84–91*
Serinus canaria, 97, 204
Setophaga ruticilla, 192, 208
sex hormones, 97
sex-reversal, 93

shearwater, **43**, 86
  Manx, 160, 161, 202
  short-tailed, 102, 202
shelduck, 164, 203
shell, 5, 107
  membrane, 6
shivering, 77, 80
shoulder girdle, 14
shrike, red-backed, 117, 135, 141, 179, 207
shrikes, 89
Sibley, 42
sight, 37, *87–89*
Sitta europaea, 61, 208
  pusilla, 166, 208
Sittidae, 208
skuas, 51
skull, 11, **46**
skylark, 127, 132, 133, 191, 206
slotted wing, 28
smell, *86–87*, 164
snipe, 22, **43**
  common, 204
soaring, 27–28
social stimulation, 166
song, 115, *126–39*
  -birds, 38, 57, 90, 91
  dialects, 135–36
  posts, 140, 193
  substitutes, 126
sound spectrograph, 128
space occupancy, 190–93
Spalding, 112
sparrow, American tree, 154, 206
  fox, 103, 206
  hedge, 191, 208
  house, 42, 69, 76, 79, 107, 116, 124, 153, 131, 198, 208
  Lincoln's, 101, 206
  savannah, 72, 206
  song, 62, 127, 137, 139, 186, 206
  tree, 199, 208
  vesper, 80, 206
  white-crowned, 98, 102, 103, 139, 153, 207
  white-throated, 146, **147**, 206
sparrow-hawk, 28, 157, 169, 189, 203
sparrows, 76, 106
speciation, 196
species, 58–64
speed, 30
Sphenisciformes, **43**, 202
Sphenodon, 20
spherical aberration, 87, 88
Spizella arborea, 154, 206
spurs, 106
starling, 69, 106, 107, 135, 140, 141, 142, 147, **151**, 157, 158, 169, **170**, 181, 184, 185, 208
stars, 158, 161, 162
starvation, 187
steamer-duck, 45, 203
Steatornis capensis, 38, 205
Steganura paradisaea, 78, **105**, 208
stereoscopic vision, 58
Sterna fuscata, 96, 204
sternum, 14, **16**, **45**
stonechat, 180, 209
stork, white, 157, **158**, 203
storks, 55
Streptopelia risoria, 197, 204
Strigiformes, **43**, 55, 205
Strigops habroptilus, 17, 205
Strix aluco, 132, 169, 205
Struthio, 46
  camelus, **48**, 202
Struthioniformes, 202
Sturnidae, 208
Sturnus vulgaris, 69, 208
subspecies, 58–64
Sula bassana, 122, 202
sun, 158, 160, 161, 162
sunbathing, **170**, 171
sunbird, splendid, 207
sunbirds, 105
supracoracoid muscle, 22, 23
supraspinate muscle, 22
swallow, **29**, 81, 89, 119, 123, 132, 133, 143, 147, **149**, 152, 154, 156, 198, 207
  barn, 207
  cliff, 118, 207
swan, **33**, **87**
swift, **16**, 22, 30, 38, 56, 81, 89, 118, 119
  Alpine, 184, 185, 205
  common, 184, 185, 205

swiftlet, white-nest, 205
swimming birds, 49–55
Sylvia atricapilla, 127, 208
  borin, 127, 208
  communis, 115, 208
  undata, 151, 208
Sylviidae, 151, 201, 208
symbiosis, 69
synscarum, 17
syrinx, **57**, 92, 126
Syrmaticus reevsii, 107, 203
Syrrhaptes paradoxus, 204

Tachyeres, 45, 203
tachypnoea, 76
Tadorna tadorna, 164, 203
Taeniopygia castanotis, 94, 208
tapeworms, 42
tarsometatarsus, 4, 18, 19
taste, 85–86
  buds, 85
taxonomy, 42–44
tegumentum, 90
temperature control, 32, *75–81*
  effects of, 153
  sense, 84
tern, sooty, 80, 96, 147, 204
terns, 89, 94
territory, *139–44*, 190
testis, 92
testosterone, 132, 139, 142
Thecodontia, 1
thrasher, brown, 132, 207
threat, 115
thrush, mistle, 69, 209
  song, **31**, 111, 120, 165, 184, 196, 209
thrushes, 127, 140, 143, 164
thyroid, 105
thyroxine, 105, 106
tibiotarsus, 18, 51
timbre, 127
Tinamidae, 49
Tinamiformes, 202
tinamou, pileated, 202
tinamous, 49
tit, bearded, 191–92, 208
  blue, 33, 110, 115, 116, **138**, 141, 143, 153, 163, 165, 167, 185, 192, 196, 207
  coal, 62, 163, 165, 167, 191, 197, 207
  crested, 165, 207
  great, 104, 110, **138**, 141, 143, 165, 167, 184, 185, 186, 188, 189, 191, 192, 196, 207
  long-tailed, 118, 120, 167, 207
  marsh, 118, 141, 165, 167, 197, 207
  plain, 141, 207
  Siberian, 165, 207
  willow, 118, 141, 165, 197, 207
tits, 85, 112, 113, 114, 118, 139, 140, 143, 150, 186, 190, 191
toes, 19, 47, 49
tone-colour, 127
tool-using, 165–66
torpidity, 81, **82**
touch, 84
Toxostoma rufum, 132, 207
trace-elements, 66
trachea, 35
tree-creeper, 167, 206
Trochili, 205
Troglodytes aedon, 197, 208
  bewicki, 197, 208
  troglodytes, 140, 208
Troglodytidae, 118, 208
Trogoniformes, 205
trogons, 205
tropical birds, 94, 102, 103, 127, 132, 145
tropic-bird, red-billed, 96, 202
  yellow-billed, 96, 202
Tucker, 61
turbulence, 26, 29
Turdidae, 127, 201, 209
Turdus merula, 69, 209
  musicus, 209
  philomelos, **31**, 209
  pilaris, 193, 209
  viscivorus, 69, 209
turkey, 20, 93, 203
Tyndall scattering, 33

Tyranni, 205
Tyto alba, **11**, **20**, 68, 205

uncinate processes, 19, 20, **21**
urea, 71
Uria aalge, **61**, 204
uric acid, 7, 9, 71

Vanellus vanellus, 28, 204
vascular system, 11–12
vertebrae, 19, **20**
Vidua, 20
paradisaea, 208
vitamins, 65, 89, 171
vitelline membrane, 6
vomer, 46
Von Haartman, 132
Vultur, 94
gryphus, 203
vulture, turkey, 86, 203
vultures, 28, 55, 86, 166

waders, 28, 33, 42, 49, 85, 89, 147, 157, 169
wagtail, African pied, 152, 207
yellow, 152, 207
wagtails, 140
warbler, Dartford, 151, 208
garden, 127, 134, 151, 196, 208
grasshopper, **128**, 208
Kirtland's, 172, 208
marsh, 135, 193, 208
reed, 134, 144, 208
sedge, 134, 144, 208
willow, 132, 133, **134**, 135, 141, 151, 154, **155**, 191, 196, 208
wood, 152, 196, 208
warblers, 151, 158, 167
water, saving of, 7, 9
Waterton, 86
waxbill, black-rumped, 73, 208
weaver-birds, weavers, 32, 44, 120, 124, 186
webbed feet, **50**, 51, **52**
Wetmore, 44, 201
whale, 49
wheatear, 33, 118, 141
whinchat, 198, 209
white of egg, 6
whitethroat, 115, 116, 159, 191, 208
whydah, paradise, 78, **105**, 107, 208
whydahs, 20, 124
wing, 13, **14**, **15**, **48**, **54**
wish-bone, 15
woodcock, **50**, 140, 204
woodpecker, great spotted, 124, 165, 189, 198, 205
green, 198, 205
lesser spotted, 167, 205
three-toed, 46, 205
woodpeckers, 38, 55, 118, 122, 126, 163
wood-pigeon, 67, 94, 101, 185, 188, 189, 198, 204
wren, **16**, 118, 132, 140, 141, 168, 171, 198, 199, 208
Bewick's, 197, 208
cactus, 80, 208
house, 80, 197, 208
wrist, 14, **15**
wryneck, 19, 179, 205
Wyre Forest, 179

yellowhammer, 132, 137, 141, 154, 206
yolk, 5–6

Zenaidura macroura, 94, 204
Zonotrichia albicollis, 146, 206
leucophrys, 98, 207
zugunruhe, 153